"However close we sometimes seem to that dark and final abyss, let no man of peace and freedom despair. For he does not stand alone. Together we shall save our planet or together we shall perish in its flames. Save it we can, and save it we must, and then shall we earn the eternal thanks of mankind."—John F. Kennedy, 35th president of the United States of America (1961-63)

Praise for Disturbing Times:

"Dis veral geskik vir omgewingsbewuste mense, studente en leerlinge in die biologie-rigting en dié wat geinteresseerd is in die toekoms van die planet."
Danila Liebenberg, Volksblad

"The text is well illustrated with pictures, graphs and statistics. It is an easy read without burdening the reader with unnecessary scientific gobbledy-gook ... Disturbing Times is an important book."
Naas Ferreira, www.artslink.co.za

"An eye-opener ... Firsing's solutions to the world's problems are fresh and alternative. Readers will enjoy his almost science-fiction-like approach to the future."
Mari Scott, Springs Advertiser

"This is one of those books you have to read."
Corinne Lamoral, Longevity magazine

"A highly interesting read."
Sunday Tribune Christmas Book Guide

Published in 2007, reprinted in 2008
by 30° South Publishers (Pty) Ltd.
28, Ninth Street, Newlands,
Johannesburg 2092, South Africa
www.30degreessouth.co.za



Design and origination by 30° South Publishers (Pty) Ltd.
Printed and bound by Pinetown Printers (Pty) Ltd.



ISBN 978-1-920143-17-6

# DISTURBING TIMES

The state of the planet and its possible future

Scott T. Firsing

30° South Publishers

# About the Author

Scott Thomas Firsing, a United States citizen, is an analyst for a risk management company based in Johannesburg, South Africa. He is a graduate of Rutgers University with a Bachelor of Science in Administration of Justice; American Military University with a Master of Arts in International Peace and Conflict Resolution; and is near completion of his doctorate in International Politics from the University of South Africa. He has multiple years' experience in various aspects of the international arena including peace, security, politics, intelligence and education. Mr. Firsing has represented multi-national corporations, government agencies, international NGOs and the United Nations. His new book, scheduled for release in 2009, is claimed to be the most extensive analysis to date of United States–South Africa relations in the post-apartheid era.

# Acknowledgments

Various people have contributed to the making of *Disturbing Times*. I would like to thank Dr. Donald Foster-Gross for taking hours upon hours out of his busy schedule to edit my book and for all his comments and suggestions. I also want to acknowledge Dr. Ken Baucom for his advice and guidance.

To my wife, Rene, for her loving support and for believing that we can make the world a better place for future generations.

# Contents

Preface 9
Audience 10
Key terms 11
Acronyms 12

**Part 1: What is the state of our world now?**

Chapter 1: The world today 16
- ❑ Introduction—the future of earth 16
- ❑ Globalization 17
  - ◆Global communications 17
  - ◆Global economy 21
  - ◆Global politics 23
  - ◆Global culture 28

Chapter 2: The ever-growing world population 32
- ❑ Feeding everyone 34
  - ◆Arable land 35
- ❑ The need for water 36
- ❑ Urbanization 39
- ❑ The world's diseases 40
- ❑ Other resource depletion 43
- ❑ Current status of alternative energy 44
- ❑ Is it ethical to have children? 46
  - ◆Alarming statistics 47

Chapter 3: The global environment 48
- ❑ Global warming 49
- ❑ Deforestation 50
- ❑ Desertification 51
- ❑ Extinction of species 52
- ❑ Our oceans 55
- ❑ Today's actions 58

Chapter 4: Protecting basic human rights 60
- ❑ Human trafficking 60
- ❑ Lack of education 63
- ❑ Corruption 66
- ❑ Migration 68
- ❑ Women's rights 72

Chapter 5: International security 74
- ❑ Terrorism 75
- ❑ Weapons of mass destruction 76
- ❑ Small arms and light weapons 78
- ❑ Today's wars 78
- ❑ Military spending 80
- ❑ Military technology 81

**Part 2: Are we moving forward?**

Chapter 6: The individual contribution 86
- ❑ Introduction—what difference can we make as individuals? 86
- ❑ Celebrities working with the UN 88
  - ◆Angelina Jolie 89
  - ◆Michael Douglas 92
- ❑ Other celebrities 93
  - ◆Oprah Winfrey 93
  - ◆Bill Gates 96
  - ◆Bono 98
  - ◆Former Presidents Bill Clinton and Jimmy Carter 100
- ❑ Lesser-known people 104
  - ◆Millard and Linda Fuller 105
  - ◆Jody Williams 106
  - ◆Clinton Hill 107
- ❑ Nobel Peace Prize 107
- ❑ Conclusion 108

Chapter 7: The contribution of international organizations 110
- ❑ Intergovernmental organizations 110
  - ◆United Nations 112
  - ◆Millennium Development Goals 115
  - ◆United Nations peacekeeping 116
  - ◆International criminal court 118
  - ◆Interpol 120
  - ◆Regional organizations 121
- ❑ Nongovernmental organizations 123
- ❑ International agreements 124
- ❑ Multinational corporations 127
- ❑ The remaining problems 131

**Part 3: Is global governance the answer?**

Chapter 8: Global governance—its history and future benefits 135

- ❑ New World Order conspiracy 136
  - ◆The Illuminati 137
- ❑ World government in entertainment 138
  - ◆*Starship Troopers* 139
  - ◆Other world government themes in entertainment 139
- ❑ World government in education 140
- ❑ Why global governance in the first place? 142
  - ◆Eliminating war 142
  - ◆Global environment 143
  - ◆Space exploration and human unity 146

Chapter 9: Global governance—how do we get there? 151
- ❑ The United States of America example 152
- ❑ The European Union example 153
- ❑ The United Nations example 155
- ❑ Important words on global governance 158
  - ◆Bill Clinton 159
  - ◆Albert Einstein 168

Chapter 10: What does the future hold? 172
- ❑ 2050—a very real scenario 173
- ❑ 2050—an alternative view 175
- ❑ Is this the generation? 178
- ❑ PeaceJam 180
- ❑ World Future Council 181
- ❑ Conclusion 184

Notes 187
Index 209
Photo credits 212

"Life on Earth is at the ever-increasing risk of being wiped out by a disaster, such as sudden global warming, nuclear war, a genetically engineered virus or other dangers we have not yet thought of."
Stephen Hawking, scientist.

# Preface

Most of us throughout the course of our everyday lives never step back and examine what is going on around us. We are so caught up on counting the days until Friday or checking our bank accounts online, that we fail to see what is occurring around the globe. For some of you the information contained within this book might sound familiar. Perhaps you read a brief headline or watched a thirty-second news clip on one of the subjects. This feeling of having read or seen something before is perfectly normal, and I promise that this sensation will come over you at least once while reading this book. Due to the vast amount of information our brain takes in on a daily basis, it needs to discard most things we encounter in everyday life. The purpose of *Disturbing Times* is not only to bring people back those discarded headlines but also to educate the reader on various other global issues. These topics are of great importance and they cannot afford to be forgotten.

*Disturbing Times* combines the latest available data from various sources. Its main goal is to prepare humanity for a global future and its potential impact on our lives. This book aims to provoke such critical thinking by presenting a realistic picture of the world that surrounds us. Although at times the facts and figures presented here might be somewhat scary, this is the true reality in which we live.

"Books are the greatest gift one person can give another."—Bono, lead singer of Irish rock band U2

# Audience

*Disturbing Times* is not targeted at one specific age group or culture but my dream is that every human being could be mindful of the material contained inside this book. Unfortunately due to the vast amount of people in this world, I realize this is simply not possible. However, above all, I encourage young people to indulge in this book and share what they have learned with their peers, because ultimately it is the younger generation who hold the future of our planet in their hands.

I have dedicated my life to making sure future generations can experience the same positive things that I have. If *Disturbing Times* influences just one teenager or young adult to dedicate his or her life to the betterment of humanity or the planet, I have succeded in doing what I meant to do when I began writing this book.

Remember, if you do find something interesting and want to research further, I strongly encourage that you do so. Awareness is the key. It is like the old saying my father use to tell me: "if we don't know where the problem lies, how are we going to fix it?" Also bear in mind that it doesn't take much to make a difference. Donating money or planting a tree is just two of thousands of things you can do as an individual to help secure a brighter future for all.

## Key Terms

In order to understand the world events that are shaping our future, it is important that we become familiar with a few key terms in world politics.

The first term you will encounter throughout this book, is the word 'actor'. By actor I do not mean a movie actor like Tom Cruise or Sean Connery. In the world arena and in this book the term 'actor' refers to the agents who are the globe's key performers. These performers are: countries like the United States (U.S.) and the United Kingdom (U.K.); international organizations like the United Nations (UN); multinational corporations (MNCs) such as Ford Motor Company and Sony; as well as nonstate actors like nongovernmental organizations (NGOs); indigenous nationalities (the Kurds in Iran, Iraq, and Turkey); and terrorist groups (the Al Qaeda global network). We will take a closer look at these different types of actors throughout the entire book.

A second term and probably the most important is 'state'. For most Americans who hear or read the word 'state', they automatically think of one of the fifty states that make up the U.S. (e.g. New York). However professors Charles Kegley and Eugene Wittkopf in their book *World Politics* define the the word 'state' as an independent, territorially defined community in the global system administered by a sovereign government and this is how we will be using the term here. For example, Australia, Canada and China are all states.

# Acronyms

**B&MGF**: The Bill & Melinda Gates Foundation
**BPI**: Bribe Payers Index

**CGI**: Clinton Global Initiative
**CIA**: Central Intelligence Agency
**$C0_2$**: Carbon Dioxide
**CPI**: Corruption Perceptions Index

**DATA**: Debt, AIDS, Trade in Africa
**DPKO**: Department of Peacekeeping Operations
**DRC**: Democratic Republic of Congo

**EC**: European Commission
**EU**: European Union

**FAO**: Food and Agriculture Organization
**FY**: Fiscal Year

**GA**: General Assembly
**GDP**: Gross Domestic Product
**GPS**: Global Positioning System
**GRACE**: Gravity Recovery and Climate Experiment

**HFHI**: Habitat for Humanity International

**IAEA**: International Atomic Energy Agency
**ICBL**: International Campaign to Ban Landmines
**ICC**: International Criminal Court
**IDP**: Internally Displaced Persons
**IGO**: Intergovernmental Organizations
**IMF**: International Monetary Fund
**Interpol**: The International Criminal Police Organization
**IPCC**: United Nations Intergovernmental Panel on Climate Change
**IUCN**: World Conservation Union

**KSE**: Kids for Saving Earth

**LRA**: Lord's Resistance Army (Uganda)

**MA**: Millennium Ecosystem Assessment
**MIT**: Massachusetts Institute of Technology
**MDARS**: Mobile Detection Assessment and Response System

**MDG**: Millennium Development Goals
**MNC**: Multinational Corporation

**NASA**: National Aeronautical and Space Agency
**NATO**: North Atlantic Treaty Organization
**NBC**: National Broadcasting Corporation
**NCB**: National Central Bureau (Interpol)
**NGO**: Nongovernmental Organization
**NRDC**: Natural Resources Defense Council

**OAS**: Organization of American States
**OECD**: Organization for Economic Cooperation and Development
**OLPC**: One Laptop per Child Association
**OPEC**: Organization of Petroleum Exporting Countries
**OSCE**: Organization for Security and Cooperation in Europe

**PEPFAR**: President's Emergency Plan for AIDS Relief
**PSA**: Public Service Announcement

**SALW**: Small Arms and Light Weapons
**SG**: Secretary-General (of the United Nations)

**THEL**: Tactical High Energy Laser
**TI**: Transparency International

**U.K.**: United Kingdom
**U.S.**: United States
**USA**: United States of America
**USAID**: United States Agency for International Development
**USDA**: United States Department of Agriculture
**UN**: United Nations
**UNDP**: United Nations Development Program
**UNESCO**: United Nations for Educational, Scientific and Cultural Organization
**UNFCCC**: United Nations Framework Convention on Climate Change
**UNFPA**: United Nations Population Fund
**UNHCR**: United Nations High Commissioner for Refugees
**UNICEF**: United Nations Children's Fund
**UNODC**: United Nations Office on Drugs and Crime
**UNSC**: United Nations Security Council

**WFC**: World Future Council
**WFP**: World Food Program
**WHO**: World Health Organization
**WMD**: Weapons of Mass Destruction
**WWF**: World Wildlife Federation

# Part 1

# What is the state of our world now?

# Chapter 1

# The world today

"Our loyalties must transcend our race, our tribe, our class, and our nation; and this means we must develop a world perspective."—Martin Luther King Jr., civil rights leader

## Introduction—the future of earth

Our universe is everything that exists. Scientists currently believe that our universe is 14 billion years old, but it could even be older if the light from objects more than 14 billion light years away has not yet had time to reach us. Our galaxy, the Milky Way, is one of billions of galaxies which make up the universe. In the Milky Way galaxy alone there are billions of stars. However, 26,000 light years from the center of the Milky Way is a small solar system with several planets orbiting around a star called the Sun.[1] Earth is the third planet from the Sun and has a total surface area of about 510 million square kilometers of which 71% of that surface is covered in water.[2]

Homo sapiens have only been on this planet for the last 300,000 to 400,000 years. You can barely compare that length of time to the history of the Earth itself, which goes back four to five billion years.[3] Nonetheless, in our short time on this planet our knowledge about the universe, the planet and ourselves has grown dramatically. In the last 100 years our advances in science and technology have led to amazing discoveries but this same knowledge has also helped damage our planet and destroy some of the species that inhabit it. Today we have amazing devices like the Hubble Space Telescope which allows us to peek further and further into space, but in turn we also possess nuclear, biological and chemical weapons that could kill hundreds of thousands of people in an instant. Unfortunately even great inventions like the automobile are leading us down a possible path of destruction. Today, wars continue, old and new diseases kill millions, and our climate is changing with serious consequences. Today, when one thinks of the future of our planet it is hard not imagine natural disasters, war, and more death.

Many people wish that this chaotic world would just go away but regrettably this is not going to happen without us taking action. Kegley and Wittkopf state, "We are all part of the world and the world is an integral part of each of us."[4] Due to this actuality, we are all influenced by events that take place all over the world.

## Globalization

"You must be the change you wish to see in the world."—Mahatma Gandhi, political and spiritual leader

The term globalization has many different definitions. In general, when people think of globalization, they mainly think about the world becoming a much smaller place. Money, supplies, ideas and even people are all moving across national borders daily and as this continues to accelerate, the world becomes more and more intertwined. In world politics, Kegley and Wittkopf define globalization as, "the integration and growing interdependence of states through their increasing contact and trade. This creates a single united global society within a single culture in order to tie people together in a common fate, thereby reducing the capacity of states to control their national destinies."[5] This definition provides a true sense of the word because it encompasses the economic, political and cultural interdependence of all states, but how did the world go from being such a large place to a global village in such a short period of time?

### Global communications

The quick advancement in science and technology is the main reason why the world shrunk and will continue to shrink. In the 1960s, when satellites

were launched into orbit around Earth, the world was slowly transformed into a global village.[6] Today, from South Africa I can instantly view and talk to my friends and family who live in the U.S. via webcam. More importantly, political and cultural events that are taking place halfway across the planet can be viewed live by people living thousands of miles away.

The main technology that made these types of actions possible was the computer. When computers first came into existence in the 1940s they took up massive amounts of space and needed full-size air conditioners to cool the vacuum tubes.[7] Now some computers are so small that they can fit in your pocket. This change was due, in part, to the space program. In order for the computers to be used in space, they had to downsize the equipment to fit in the tiny capsules.[8] Today computers are found in everything from microwaves to automobiles.

This growth of computers and the miniaturizing of the chips that drive them have not only transformed our jobs but have also benefited various other aspects of life. One of the more important fields that has changed considerably has been the health sector. What started with more powerful microscopes and ultra sound has evolved into MRIs and CAT scans.[9] The newer fields include genetic engineering and nanotechnology. All of these breakthroughs wouldn't have been possible if it wasn't for the advent of the computer.

To many, the advantages of the communications revolution are, through globalization, a blessing for humanity. In a report from the UN Development Program (UNDP) it states, "as the communications revolution turns digital promises .... to connect everything to everything else, are creating ... a tremendous force for human development for all those connected—by providing information, enabling empowerment, and raising productivity."[10] The Internet can even threaten government's rule, like China needing to block various websites from its citizens.

With all the benefits it is hard to see the downside in the growth of global communication. Nonetheless, that side is there. Some feel that the communications revolution is widening the gap between the rich and the poor, leading to the globalization of poverty and the destruction of local culture.[11]

The November 2005 report titled 'Toward Knowledge Societies' by the United Nations for Educational, Scientific and Cultural Organization (UNESCO)

stated that nearly half of the 6,000 languages spoken today on Earth run the risk of disappearing this century. They found that this could be due to the use of new technologies such as the Internet. "Three of each four pages on the Internet are written in English. Nevertheless, the number of web surfers whose mother tongue is not English exceeds 50%, a percentage that continues enlarging," states the document.[12]

The report also discussed what is known as the 'digital divide'. As the communication revolution continues, it is seen as only benefiting a few countries while putting the rest at a disadvantage. As Kegley and Wittkopf point out, "The doors to the communication revolution for small entrepreneurs in developing countries remain closed, creating a digital divide where one third of the world's six billion people are denied access."[13] The UNESCO document urges countries to favor universal access to information technologies and communication, to reduce the existing gap between the rich and poor countries. Only 11% of the world's population has Internet access and 90% of those users live in industrialized countries.[14]

UNESCO believes the answer to the problem may be to spread and use free software and accessible data-processing teams in developing countries while also creating more public centers that have Internet access. The document supports universities, museums, libraries and other centers because they permit random access to all information that is in the public domain.[15]

Another idea has been the US$100 laptop. Former UN Secretary-General Kofi Annan unveiled the first working prototype of the US$100 laptop on November 16, 2005 at the World Summit on the Information Society in Tunis, Tunisia. Annan was joined by Nicholas Negroponte, chairman and co-founder of the Media Lab at MIT, in presenting the laptop to the gathering.[16]

The US$100 laptop is an ultra-low-cost, full-featured computer designed to significantly enhance children's primary and secondary education worldwide. This idea is the main project of the nonprofit One Laptop per Child (OLPC) association. Their aim is to equip all the world's schoolchildren and their teachers with a personal, portable, connected computer.

Kofi Annan stated: "The $100 laptop is inspiring in many respects. It is an impressive technical achievement, able to do almost everything that larger, more expensive computers can do. It holds the promise of major advances in economic and social development. But perhaps most important is the true meaning of 'one laptop per child'. This is not just a matter of giving a laptop to each child, as if bestowing on them some magical charm. The magic lies within—within each child, within each scientist, scholar, or just plain citizen-in-the-making. This initiative is meant to bring it forth into the light of day."[17]

Mr. Negroponte added that, "Children are the greatest natural resource of any country and educating these children is at the root of solving our largest and most complex problems. Yet the best education may not come from sitting in a traditional classroom, but rather through independent interaction and exploration. The development of a $100 laptop will now make this possible for all kids—especially those in developing nations. It will redefine how we 'learn learning'."[18]

These laptops will be sold to governments and issued to children by schools on a basis of one laptop per child. The laptops will be so energy-efficient that hand-cranking alone will generate plenty of power for operation. It also has mesh networking, which will give many machines Internet access from one connection. Although the price is US$100 now, it is expected that it will decrease steadily over the next few months and years.

The World Summit on the Information Society is the result of three years of planning. Its goal is to put the spotlight on ways to bridge the digital divide and harness the power of the new information and communication technologies to spur progress towards the UN's Millennium Development Goals (*see* Chapter 7). In attendance at the 2005 meeting were more than 45 heads of state and delegations from more than 150 nations, with a total of 16,000 people attending.[19]

**Fixed lines and mobiles for every 1,000 people**

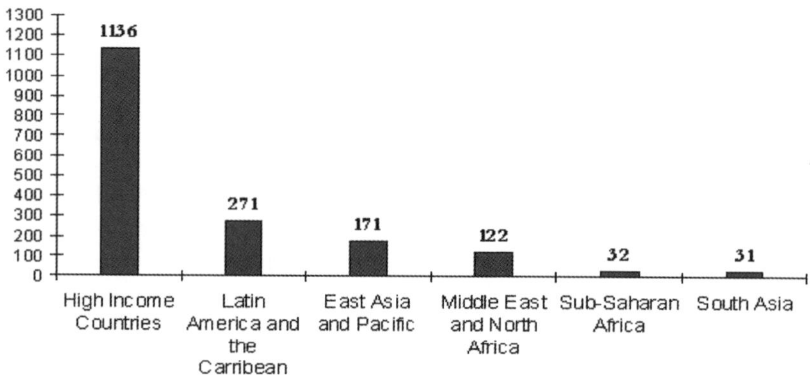


(Source: World Bank)

**Ownership of PCs for every 1,000 people**

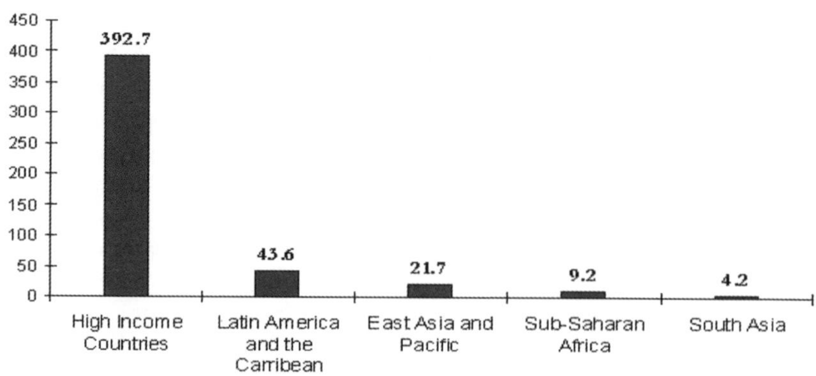


(Source: UN Statistics Division (ITU))

**Total number of Internet users (in millions) and % total Internet use**

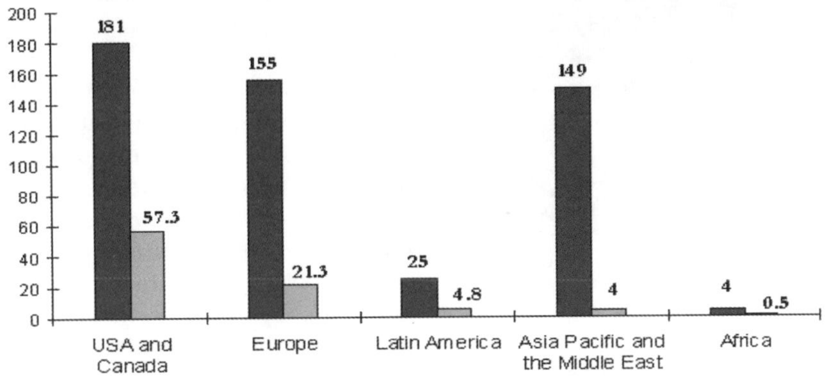


(Source: nva.ie.dprb.org)

## Global economy

Economic globalization is also a very important and widely debated topic in the world today. The global economy includes all the financial interactions among businesses, people and governments. This growth has been due not only to the communications revolution but also to advances in sea and air transportation. Every day huge cargo ships inexpensively carry all sorts of products from one part of the world to the other. This together with quick, easy and secure global transactions, has linked economies, corporations and individuals to all parts of the Earth.

Multinational Corporations (MNCs) such as ExxonMobil, British Petroleum and Toyota produce components of their products on different continents and may assemble them in various locations. This global economic integration allows such companies to view the whole world as the market for their goods.[20] Like most aspects of globalization, people see both positives and negatives to MNCs (*see* Chapter 7).

People against economic globalization have the same claim of those against the communication revolution; that globalization is making the rich richer and the poor poorer. Those who question the benefits of globalization point out that poverty has remained stubbornly high in sub-Saharan Africa. Between 1981 and 2001 the fraction of Africans living below the international poverty

Gateway to Mumbai. The city of Mumbai is home to some India's brightest tech minds.

line increased from 42 to 47%.[21] The integration into the international economy can be very difficult. Most poor countries provide very little effective protection to help people who have lost their jobs. Unlike the U.S. and the U.K., when these people lose their jobs and cannot find new ones they don't have unemployment money from the government to fall back on. Often people end up begging as they don't have money for food.

Professor of Economics at UC Berkeley, Pranab Bardhan, points out that, "Small exporters in poor nations often lack the marketing networks and brand names to make inroads into rich-country markets. Although transnational retail companies can help them, the margins and fees they charge are often very high."[22] Bardhan gives the evidence of the international coffee market which is dominated by four companies. In the early 1990s the coffee earnings of exporting countries were US$12 billion with retail sales of US$30 billion. "By 2002, retail sales had more than doubled, yet coffee-producing countries received about half their earnings of a decade earlier. The problem is not global markets but impeded access to those markets or depressed prices received by producers, as a result of the near-monopoly power enjoyed by a few retail firms."[23] In addition, in certain industries, companies may actively conspire together to fix prices.

Many researchers have found that if a particular country's government and economic policies are not open to global trade that the country and its people will continue to be worse off. Contrasting case studies of countries prove this. Although the island economies of Mauritius and Jamaica had similar per capita incomes in the early 1980s, their economic performance since then is dramatically different. Mauritius has better participatory institutions and its citizens obey the rule of law, while Jamaica is full of crime and violence.[24] Another example is Botswana and Angola. These two are both diamond-exporting countries in southern Africa but Botswana is democratic and fast-growing, while Angola is in and out of civil war.

The growth of the global economy has also had a major impact on the use of energy and other resources which in turn has led to political and environmental problems.[25] Due to the fact that manufacturing requires raw materials, many states rely on them for development. In the 20th and now 21st century, oil has been and will continue to be one of those main resources.

Oil reserves have become so important that the states that possess oil have the ability to affect other states' economics and their political stance. Some of us are too young to remember but in the 1970s the Organization of Petroleum Exporting Countries (OPEC) declared an oil embargo restricting trade. My father told me that during the embargo he had to wait for hours for gasoline and that some days gas stations did not have any petrol at all. This forced a fairly large economic decline for many countries during that time period.[26] Today we see the impact the price of a barrel of Brent crude oil can have on economies all over the world.

The first Iraq war also showed us the true significance of oil. When Iraq, under the dictatorship of Saddam Hussein, invaded Kuwait in 1990, people began to think that he would also invade neighboring Saudi Arabia. The world's countries were terrified because if Saddam took over both countries he would have control of almost all of the world's petroleum supplies. The UN tried economic sanctions to no avail. Then in early 1991, a coalition of some 39 nations declared war on Iraq.[27] The Persian Gulf War shows how dependent we are on oil and how all of countries are linked economically.

Not only has the growth of the global economy had political implications, it is having a major impact on the environment. When resources like oil and coal are burned carbon dioxide is released in the atmosphere. This causes pollution, acid rain and damages the ozone. Chapter 3 provides a complete analysis of the environmental problems that Earth now faces.

### Global politics

In the past few decades world politics has changed substantially. There has been a large growth in international and regional organizations (*see* Chapter 7). However, looking at the individual states around the world we see the growing trend of democracy. Some theorists have long maintained that the types of governments in the world are relevant to the probability of armed conflict. Democracies, they insist, rarely (if ever) go to war with one another.

U.S. President Bush giving a speech on Darfur at the UN in New York.

Since the mid-1970s, the global growth of democratic governance has skyrocketed. Kegley and Wittkopf found that since the mid-1970s, "freedom and civil liberties have grown, in a series of transitions, and today the enlarged community of liberal democracies make up three-quarters of the globe's governments, ruled by free or partially free institutions."[28] What has caused a boom so big?

A series of clear factors have propelled the democratization movement in the world. First, the justifications offered historically as defenses of authoritarian rule (for example, strong central government is necessary to quell ethnic, religious, or regional tensions) are no longer popularly accepted.[29] Second, the increased power of the electronic and print media has made information available to all segments of populations, and in turn has inspired more democratically oriented or mass-oriented organizations to mobilize. Lastly, most external donors and organizations now either require a country to be democratic or in the process of, before they will commit to assisting that country or any other country in need of help.

The last reason listed is by far the most important. From a strictly economic viewpoint, organizations such as the World Bank, the International Monetary Fund (IMF), and the Organization for Economic Cooperation and Development (OECD), all endorse the promotion of democracy as a policy priority.[30] This has forced many countries to initiate serious policy reform so they can show they meet those criteria of good governance. Although the governance criteria are based on Western concerns for freedom and human rights like speech, press and so forth, the World Bank's standards which must be met are accountability, transparency, predictability, the rule of law in government, and the absence of any factors that could interfere with economic recovery.[31] Looking at the Economic Commission for Africa, nations must have a government guided by what Dr. Adebayo Adedji, (Former Executive Secretary of the Commission) calls the 'five Cs'. They are consensus, commitment, conviction, compassion and compromise.[32] Again, the importance of assistance from the international donor shows us why so many countries around the world have taken on democracy.

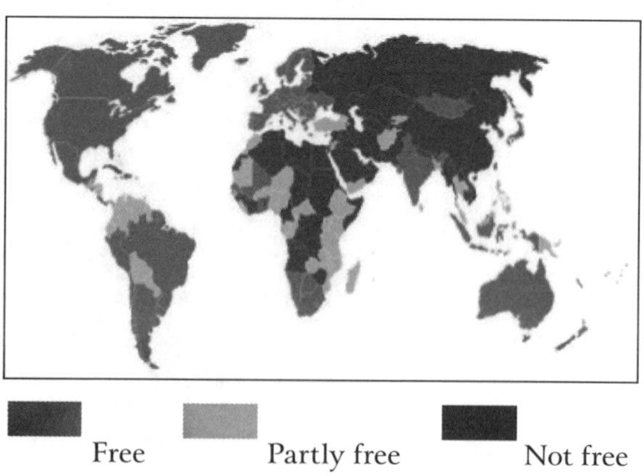

**The trend toward freedom (1973–2003)**

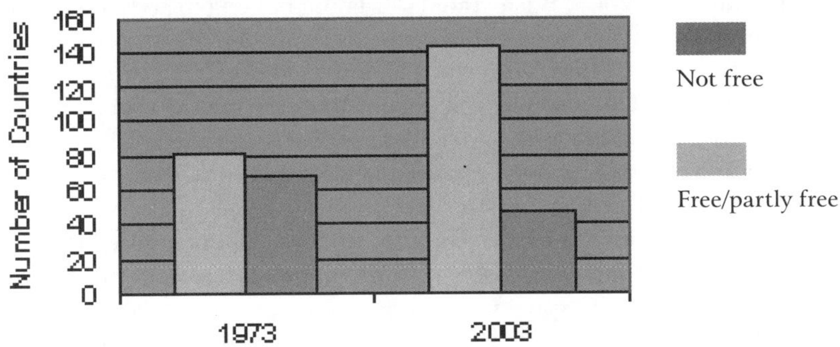


A growing number of free nations

(Source: Freedom House)

The map (page 24) from Freedom House's survey, Freedom in the World 2006, reports the state of world freedom in 2005. It is one of the most widely used measures of democracy by researchers.

Other organizations such as the UN are also key actors in spreading democracy. They are required to perform a wide range of roles and services in the interests of restoring and establishing democracy and good governance practices. Some of these activities are considered in Boutros Boutros-Ghali's *Agenda for Peace,* which the former UN Secretary-General defines "peacebuilding" as a broad set of activities that "tend to consolidate the peace and advance a sense of confidence of confidence and wellbeing among the people ... [The list of peacebuilding activities includes] disarming the previously warring parties and the restoration of order, the custody and possible destruction of weapons, repatriating refugees, advisory and training support for security personnel, monitoring elections, advancing efforts to protect human rights, reforming or strengthening governmental institutions and promoting formal and informal processes of political participation."[33]

Another major push to spread democracy in the international system comes from the U.S. In the 1980s and 90s, both Presidents George H. W. Bush and Bill Clinton proclaimed the promotion of democracy a continuing foreign policy priority of the United States after the Cold War. Bush in his famous 'New World Order' speech included global democratization among the key elements. Bill Clinton declared the "enlargement of the democratic community" to be the centerpiece of his foreign policy.[34] Both presidents cited the complementary relationship between global economic integration and political reforms. They both believed that a world made up of democratic states would be more peaceful. Top officials in the State Department shared these views as well, and put the policy, labeled 'Building Democracy', into practise.[35]

Today, George W. Bush shares the same views as his father and Mr. Clinton. In his inaugural address after being sworn in for his second term in office in January of 2005, Bush stated, "It is the policy of the United States to seek and support the growth of democratic movements and institutions in every nation and culture, with the ultimate goal of ending tyranny in our world."[36] The President also pledged to continue efforts to spread democracy throughout the world. He said, "We will persistently clarify the choice before every ruler and every nation—the moral choice between oppression, which is always wrong, and freedom, which is eternally right."[37]

A year earlier, on February 10, 2004, Paula Dobriansky, the U.S. Undersecretary of State for Global Affairs, addressed a Baltimore foreign affairs group, calling the spread of democracy the essential part of the war on terrorism. She stated, "It is our belief that the spread of liberal democracy is an essential part of a long-term strategy to win that war."[38] Dobriansky also named several international projects to strengthen democracy that the U.S. is supporting. This list includes the Community of Democracies, "a new international network to which young and old democracies of different size and form, gather to strengthen representative government, to share experiences, to help one another, and to coordinate policies in areas of common interest." Another organization called the National Endowment for Democracy is a private, non-profit organization supported predominantly by the U.S. Government that directs grants among other things. Lastly, the Millennium Challenge Account, a new U.S foreign aid program designed "to reward the countries that work the hardest for the interests of their people ... that are committed to governing justly, investing their own resources in their people, and working to provide a proper climate for economic growth." She concluded by saying that the long-term strategy, "is to strike at the heart of terrorism by depriving it of its havens, its recruitment grounds, and its foot soldiers. We will do this in no small measure through the avid promotion of democracy and freedom."[39]

In addition to supporting these particular organizations, the U.S. promotes democracy through their Agency for International Development or USAID. To get a better feel and a more detailed look into what exactly the United States is doing and where the money is going, we can examine one particular country, in this case South Africa. For the fiscal year of 2006, the U.S. gave a total of about 35 million dollars, down from 62 million in 2004.[40] This money is broken down into six sections: democracy and governance; education; economic capacity building; housing and municipal services; HIV/AIDS and primary health care; and employment creation.

U.S. Embassy in Pretoria, South Africa.

Although it sounds simple enough

for countries to change to a democracy and end conflict, the world today still faces many issues. One major problem is that the majority of people in several U.S.-allied countries do not believe it is the U.S.'s role to spread democracy around the world. The AP-Ipsos poll surveyed about 1,000 adults in nine countries which included Britain, France, Germany, Italy, Spain, Canada, Mexico, South Korea and the United States. The poll found that "84% of those surveyed in France believe it should not be the U.S.'s role to spread democracy in the world, and 78% of Germans and two-thirds of Britons also rejected the idea. A majority of people in Italy, Spain, Canada, Mexico and South Korea were also opposed. Roughly 53% of Americans say the United States should not be trying to spread democracy abroad, while 43% believe their country should do so."[41] This seems consistent with Bush's go-at-it-alone strategy, but trying to spread democracy all by yourself not only seems impossible, but it may also be a reason why we are facing the war on terror in the first place.

Historian and author Eric Hobsbawm believes that although the U.S. is adjusting to the ever- present threat of terrorism, they are slowly waking to the idea that some things are simply beyond its control. "Democracy has kept things running fairly smoothly at home, but outside the white picket fence that surrounds its borders, America faces a league of nations that is increasingly unwilling to accept its hegemonic view."[42] Hobsbawm adds that the assumption by the U.S. that countries can and will be remade in its image is one that ignores the complexities and pitfalls inherent in such a radical shift. Furthermore, "A state cannot practice effective democracy without consensus, for example, as citizens in countries such as Sri Lanka, Czechoslovakia, and Northern Ireland can attest. In fact, oftenzztimes a people's mandate in countries riddled with social and economic strife can lead to systematic dysfunction, corruption, and even fascism."[43]

Many share the view that the democratization process and conflict go hand in hand. Professors Edward D. Mansfield and Jack L. Snyder in Michael Brown's collection of *Theories of War and Peace* found that in this transitional phase of democratization, countries become more aggressive and war-prone, not less, and they do fight wars with democratic states. They completed their study using the same databases typically used to study the democratic peace. "Moreover, virtually every great power had gone on the warpath during the initial phase of its entry into the era of mass politics."[44] Mansfield and Snyder use examples such as Japan's 'Tiasho democracy' of the 1920s, which brought an era of mass politics that led the Japanese army to devise and sell an imperial ideology with broad-based appeal. But why is this the case?

Mansfield and Snyder found that these new democracies are characterized by weakly institutionalized democratic procedures, and they often confront illiberal and radical social groups. In addition, "elites in democratizing states cannot easily form stable conditions or pursue coherent policies, so that appeals to nationalism and risky foreign policies become attractive ways to shore up domestic political power."[45]

University of Maryland professor, Dr. Ted Gurr explains that the most dubious expectation of all is to believe that authoritarian states such as Sudan, Burma, and Burundi may be able to defuse ethno-political wars by moving toward democracy. "Democratic institutions in societies driven by ethnic rivalries are more likely to increase incentives and opportunities for more fighting than to provide pathways to peaceful accommodation."[46] Gurr also found that new democratic regimes usually lack the resources or institutional means to make and guarantee the kinds of accommodations that typify the established democracies. "Therefore, democratization in its early stages facilitates both ethnically based protest and rebellion."[47]

Federal Bureau of Investigation (FBI) photo of Al Qaeda Leader Osama bin Laden.

Many peace processes in the 1990s went awry and today the spread of democracy throughout the globe continues to be a long, hard-fought battle. "In the long run the enlargement of the zone of stable democracy will probably enhance prospects for peace. In the short run much work remains to be done to minimize the dangers of the turbulent transition."[48] The international community will play a large part in this. It is on their shoulders to promote democracy and to maintain peace in today's multiethnic societies.

We cannot allow terrorists like Osama bin Laden who believe they can change U.S. policy through terror, to force democratic states to suspend their civil liberties.[49] Bush states, "... In the long run, there is no justice without freedom, and there can be no human rights without human liberty."[50] Again, this task of a global democracy will not be easy. Prime Minister Ingvar Caarlsson of Sweden might have said it best when he warned, "If we fail to nurture democracy—the most fundamental political project of this century—we will never be able to realize our goals and the responsibilities which the future will call for."[51]

**Global culture**

Globalization has also affected popular culture which includes everything from music, movies and sports to food, hobbies and fashion. This change is also due to science and technology. With the invention of satellite television and the Internet, mass media now has the ability to reach billions of people daily. This is true even in the emerging world. Recent estimates suggest that about half the population or some 2.5 billion people have regular access to a television set.[52]

McDonald's outlet in Shanghai.

Television, radio and movies are among the world's most popular forms of entertainment. These media not only entertain but also show people how other people live in other parts of the world. The West dominates the worldwide mass media so it is no surprise that they have influenced many different cultures around the globe. Beck *v.* give many examples of western influence. For instance, "blue jeans are the clothes of choice of most of the world's youth. Western business suits are standard uniforms among many people. American-style hamburgers and soft drinks can be purchased in many countries around the world. Mickey Mouse and other Disney characters are almost universally recognized."[53] This influence has also changed people's thinking with the best example being the many different cultures that have adopted materialism or the Western mindset of placing a high value of acquiring material possessions.

Flag burning in Tehran, Iran.

**Jihad vs. McWorld**

In 1995 Benjamin Barber published his book *Jihad vs. McWorld*. By using the word McWorld, Barber portrays a playful image of the hamburger giant which was among one of the first MNCs. This force in the world is primarily economic while Jihad on the other hand is political, as it is the Islamic term for holy war. Both 'McWorld' and 'Jihad' are in fact cultural forces in that their power is derived from their capacity to publicize images that hold the world's attention and however different they might appear to be, they are called into being by the same process of economic globalization.[54]

Overall, Barber believes that the spreading of Western ideas or companies shows citizens in smaller nations that the West is truly at the peak of its power. While the interactions between civilizations are increasing globally, those interactions intensify the consciousness of one's own civilization locally. In other words, the spread of McDonald's and other MNCs might bring about a stronger bond to local groups which could lead to terrorist organizations. As Barber would say, the spread of McWorld intensifies Jihad.

Barry Rubin, Director of the Global Research in International Affairs (GLORIA) Center of the Interdisciplinary Center, wrote an article explaining why globalization has had a different effect in the Middle East. He explains that the majority of regimes, opposition movements, and intellectuals in the region are consciously anti-globalization. "Moreover, there is no part of the world where violence is more often used in the anti-globalization struggle, most notably by Osama bin Laden but generally by all radical Islamist movements."[55]

Mr. Rubin also points out that it is important to remember that Islam is a religion with its own set of laws and claims to provide the proper order for society. Thus, many elements of globalization can contradict—or be thought to conflict with—Islam, which is far less the case with other religions like Buddhism and Hinduism.

One of these elements is the use of the English language around the world. Rubin states: "Rejection of Western culture can also be seen through the limited use of European languages in the Middle East. English, while empowered by being the major language of global interchange, has been very much restricted. French has declined sharply as an important language in North Africa, and other European languages are almost totally absent. The principal language of all the Islamic Middle East is Arabic, except in Iran, and Arabic is also the language of Islam. The existence of a large and culturally powerful Arabic community builds a linguistic wall against the penetration of Western languages which are such an important aspect of globalization."[56]

Due to factors like language, among others, the basic elements of globalization are seen as more alien in the Middle East than elsewhere, and are thus far more likely to be seen as hostile. Latin American countries might worry about being overwhelmed by different food or music, but the Islamic Middle East is likely to look at the entire list of globalization ideas and products as dangerous.[57]

This hostility against globalization and the West can be seen on satellite television, most notably al-Jazira in Qatar. They have been known to spread extremist doctrine most effectively. Rubin writes, "Al-Jazira programs often feature a debate between a modernist and a militant nationalist or Islamist. The program's host makes it clear that the latter side is correct and the former argument is unacceptable or even evil. Screened telephone calls then support the radical side without exception."

There are countries of the world where the USA flag desecration incidents illustrated in the websites photo archive took place. It is easy to spot the countries not represented in the archive (black), which are primarily in Africa and central Asia, as well as some others, including North Korea, Myanmar, Cuba, and Papua New Guinea.

Students who have studied in the West more often than not return home to reinforce even further a rejection of the society they have just experienced.[58]

Having lived in a Western country, they have seen how the society operates and they may focus on its faults. Additionally, they might fear the effects of such ideas or institutions if introduced in their own countries. If these individuals feel tempted by the things they have experienced in the Western countries, they may feel guilty and thus want to reinforce their own national and religious character. In fact, "many of the September 11 hijackers had experience of living in the West, reinterpreting Islam in a way that focused on making it a shield against any Western or globalization's influence and a sword to destroy those things altogether."[59]

The spread of culture such as books, cars, and clothes has only just begun and because of this, it will take several years to see how influential cultural globalization really is. But do not forget as Barry Rubin highlights, "the depth of Middle Eastern opposition to globalization should not be underestimated. Every indication is that it will extend years, even decades, into the future."

To conclude this chapter, at this moment in time all of the world's states are linked politically, economically and culturally. There is fear and uncertainly that goes along with this. Although globalization has been a blessing in many ways, it has also led to many serious problems we need to address. In this chapter we briefly touched on several of these particular issues like poverty, terrorism, and the environment. However, there are also several other issues that may have nothing to with globalization that need to be examined more closely along with the ones mentioned previously. In the next few chapters we will take a closer look at all of these topics starting with the ever-growing world population.

# Chapter 2

# The ever-growing world population

"We cannot control the massive challenges of poverty, hunger, disease, and environmental destruction unless we address the issues of population and reproductive health."
—Thoraya Ahmed Obaid, Executive Director, UNFPA

Overpopulation has become a very big topic in world politics. The world population began to accelerate after WWII when drugs such as penicillin were able to reduce the death rates, causing 'a population explosion'. According to the U.S. Bureau of the Census, the world population on June 1, 2007 will be 6,598,721,723 people. By 2050, world population is projected to reach nine billion people.[1] That is more than five times the number of people alive in 1900. Most of this growth will come from the developing world while most of the industrialized nations will see decline in births and thus more aged populations.[2]

According to Michael Bowman, a reporter for Voice of America, by 2050 Africa's population, both northern and sub-Saharan, is expected to surge from 900 million to almost two billion, while South Asia's population is projected to swell from 1.6 billion to nearly 2.5 billion. At the same time, Europe's population is expected to shrink from 730 million to 660 million.[3] As you can see, if these projections hold true most of the future growth will be concentrated in the Africa, South Asia, and Latin America. Carl Haub, a senior demographer at the Washington-based Population Reference Bureau states, "All world population growth today is in the developing world. There is no natural population growth in Europe, and even the U.S. is very heavily dependent on immigration."[4]

It is hard to picture that every year more than 81 million people enter this world, and every ten years almost one billion inhabitants are added to the world's population. How can the number of human beings grow so quickly?

Kegley and Wittkopf use astronomer Carl Sagan's principle on growth rates which he calls the 'The Secret of the Persian Chessboard':

"The way I first heard the story, it happened in ancient Persia. But it may have been India, or even China. Anyway, it happened a long time ago. The Grand Vizier, the principal adviser to the king, had invented a new game. It was played with moving pieces on a board of 64 squares. The most important piece was the king. The next most important piece was the Grand Vizier—just what we might expect of a game invented by a Grand Vizier. The object of the game was to capture the enemy king, and so the game was called, in Persian, shahmat—shah for king, mat for dead. Death to the king. In Russia it is called shakhmaty, which perhaps conveys a lingering revolutionary ardor. Even in English there is an echo of the name—the final move is called 'checkmate'. The game, of course, is chess.

As time passed, the pieces, their moves and the rules evolved. There is, for example, no longer a piece called the Grand Vizier—it has become transmogrified into a queen, with much more formidable powers.

Why a king should delight in the creation of a game called 'Death to the King' is a mystery. But, the story goes, he was so pleased that he asked the Grand Vizier to name his own reward for such a splendid invention. The Grand Vizier has his answer ready: He was a humble man, he told the king. He wished only for a humble reward. Gesturing to the eight columns and eight rows of squares on the board he devised, he asked that he be given a single grain of wheat on the first square, twice that on the second square, twice that on the third, and so on, until each square had its complement of wheat.

No the king remonstrated. This is too modest a prize for so important an invention. He offered jewels, dancing girls, palaces. But the Grand Vizier, his eyes becoming lowered, refused them all. It was little piles of wheat he wanted. So, secretly marveling at the unselfishness of his counselor, the king graciously consented.

When the Master of the Royal Granary began to count out the grains, however, the king was in for a rude surprise. The number of grains starts small enough: 1, 2, 4, 8, 32, 64, 128, 256, 512, 1,024 ... but by the time the 64th square is approached, the number becomes colossal, staggering. In fact the number is nearly 18.5 quintillion grains of wheat. Maybe the Grand Vizier was on a high-fiber diet.

How much does 18.5 quintillion grains of wheat weigh? If each grain were two millimeters in size, then all the grains together would weigh around 75 billion metric tons, which far exceeds what could have been stored in the king's granaries. In fact, this is the equivalent of about 150 years of the world's present wheat production."[5]

## Feeding everyone

With all of the people in this world, will it be possible to feed everyone? On the one side of the argument are the optimists who believe that due to factors such as genetic engineering and transgenetic crops, there will always be enough food to go around. Genetically engineered transgenetic crops are altered by inserting genes to develop a desired trait, such as an herbicide tolerance or increased oil content.[6] On the other side of the argument are the pessimists.

In the article entitled, 'Grain Production Dropping; Fuel Thefts Rising'. by Ralph Woodgate of World Population Awareness, he states: "Many years ago while working in India, on one of our later trips, I was asked to meet with a senior government official in New Delhi. I had no idea what he wanted, in fact I had only briefly met him once before. 'You have been coming to

India for several years,' he said. 'I would like to hear your comments, as an impartial observer, on the changes you see in our country.' I told him that when we first came to India we were horrified at the conditions we saw. With the years things had dramatically changed, and we no longer saw the signs of starvation that had concerned us in the past. It appeared that the standard of living for the whole country had improved dramatically. I told him that in my opinion they should be congratulated for all that had been done for the people. I will never forget his response. He smiled and thanked me, but then he shook his head and said, 'We have worked very hard to improve our agriculture and now we can feed all our people, and even export a small amount of food. But next year we will have a million more mouths to feed, and the year following a further million and the next year it will be five millions more. We cannot keep up with the population growth and if it continues we face the worst famine the world has ever seen.'"[7]

In 2006 the world's grain harvest was 61 million tons short of consumption. Moreover, production has failed to meet demand in six of the last seven years.[8] The U.S. Department of Agriculture (USDA) estimates this year's global grain harvest at 1,984 million tons, will be down 24 million tons from last year. It is 3% below the high of 2,044 million tons in 2004. World grain consumption has risen every year in the last 45 years, except for 1974, 1988 and 1995, when sharp price hikes and tight supplies pushed consumption down.[9]

This demand for grain is driven largely by population growth. Today there is also the issue of food vs. fuel, as grain is now being grown for grain-based ethanol for cars and trucks. The current statistics are 60% of grain is used as food, 36% as feed, and 3% as fuel, however, the fuel percentage is predicted to grow by more than 20% each year. This will only continue to get worse as the world's farmers must try to feed another 70 million people each year, regardless of weather.[10]

Looking at other sources of food, improving incomes are allowing three to four billion consumers to eat more poultry, pork, beef, milk, eggs and farmed fish. The article states that, "Global meat production leaped from 44 million tons in 1950 to 265 million tons in 2005 and keeps climbing. The U.S. is expected to use 20% of the projected harvest to produce corn ethanol. That consumption will match our export and exceed Canada's total harvest."[11] Other sources have even higher numbers.

**Arable land**

According to the UN Food and Agricultural Organization (FAO), human activities linked to population pressures and unsustainable agricultural practices, have destroyed or severely degraded 11% of the world's arable land. This is an area roughly the size of India and China combined. As a result, every year the world's farmers must feed tens of millions more people with 27 million fewer tons of topsoil.[12]

Africa may have the biggest problem regarding land for food. Farm holdings

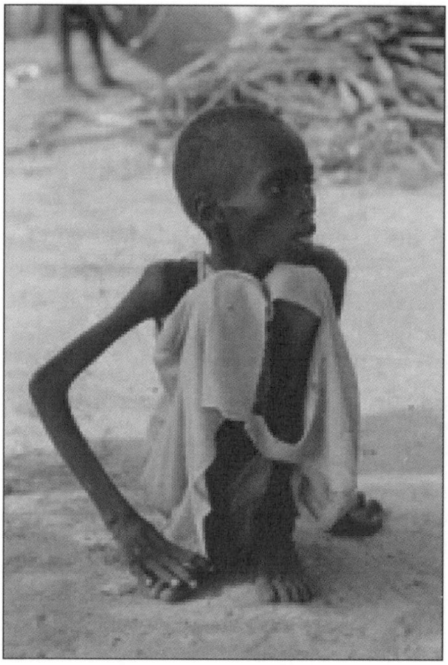

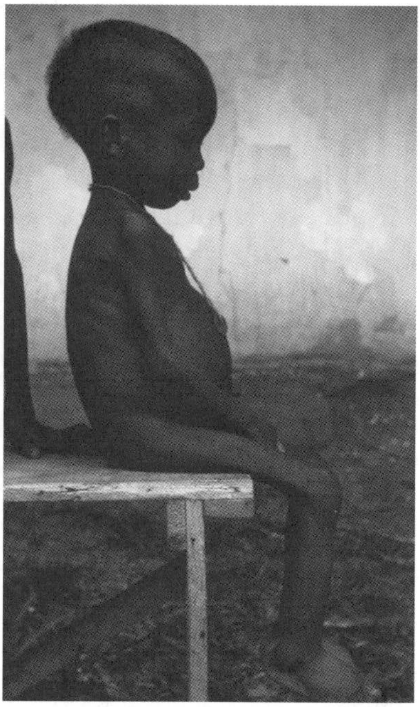

have shrunk as the population has grown, and per capita yields have fallen by as much as 30% over the last three decades.[13] Some 65% of the region's agricultural land has been degraded. FAO writes, "As land use intensifies, fallow periods decline and cultivation spreads into marginal and ecologically fragile lands."

The FAO states that while Africa's population grew by 3% a year in the three decades to 1996, its annual food production increased by only 1.9%.[14] Due to these figures, the number of malnourished children in Africa could grow from 29 to 41 million between 1980 and 2020—a prospect not helped by the spread of HIV/AIDS.

Africa, of course, is not alone in facing population-related pressures on the land. China, with one of the biggest population growths, has lost arable land equivalent to all the cropland in Denmark, France, Germany and the Netherlands combined. High population growth and low yields have forced millions of small farmers to clear forests, overgraze and cultivate marginal lands, causing soil erosion and deepening rural poverty.[15] This, along with skewed land holdings (especially in Latin America), unsustainable logging, ranching, mining and plantation farming has added to the problems of environmental sustainability (*see* Chapter 3). All in all, the UN World Food Program (WFP) classifies 88 states as unable to provide enough food for its people.[16]

## The need for water

Another problem that the whole world faces is increasing water shortages. As the population continues to grow, the world's fresh water supply does not

Picture of Korat, Thailand. Waters in Lam Takong Dam in Korat, Nakhon Ratchasima province of Thailand have dried up due to prolonged drought, allowing villagers to camp inside the dam to catch the remaining fish. Severe water shortage and damage to agriculture has affected millions.

increase with it. The UN states that 2.7 billion people will face severe water shortages by 2025 if consumption continues at current rates. Unfortunately of all the water on the planet, 97% is salt water from seas and oceans, 2% is frozen in polar ice sheets and glaciers, and a fraction of 1% is for drinking, irrigation, and industrial use.[17]

A report sponsored by the FAO and the Consultative Group on International Agricultural Research entitled 'Water for Good, Water for Life: Insights from the Comprehensive Assessment of Water Management in Agriculture', found that parts of the world, including the breadbaskets of India and China, the cotton belt of Central Asia and swathes of the Middle East, are reaching the physical limits of their water supplies. The world's poorest region, Sub-Saharan Africa, also has a major problem regarding water. They currently lack the money to build dams and irrigation systems to get water to farms and homes in the rural areas where most people live.[18] David Molden, the principal researcher at the International Water Management Institute in Sri Lanka and the coordinator of the study states, "We have to learn how to grow more food with less water. That's imperative. We can't just keep expanding the land used."[19]There are simple technologies that would help many developing countries with their situation, but the real problem lies with implementation. The authors of the study note that while these technologies may be simple, installing them on a national scale and maintaining their use would be no easy matter. For example, a country like Ethiopia, with very low rural literacy levels, would need to train people to carry out such a plan. Furthermore, just having more water will not be enough. "Most farmers need credit, crop insurance and roads to get their products to market. They need AIDS treatment, and they need fertilizers to nourish their land. A major study released in March of 2006 found that three-quarters of sub-Saharan Africa's farmland is severely depleted of basic nutrients to grow crops."[20]

Another report by the World Wildlife Federation (WWF) says that a combination of climate change and poor resource management is leading to water shortages in even the most developed countries. It urges water conservation on a global scale and asks for rich states to set an example by repairing their ageing water infrastructure and by tackling pollution. The WWF states that in London, leaks from ageing water mains are wasting 300 Olympic swimming pools' worth of water every single day.[21] *National Geographic* found a similar situation in Mexico City. There, the population is booming while the water pipes crumble and because of it, the city must truck in water for many of its residents. Mexico City was once called the Venice of the New World for what are now long-gone lakes and canals. The city has drained its aquifer since 1900 and has even sunk 24 feet.[22] As this ground shifts, the water pipes break. Water leaks now claim nearly a third of its water.

Dubai, United Arab Emirates—The Rare Example. *National Geographic* states that in bone-dry Dubai 2.6 million gallons of expensively desalinated seawater flow through places like water parks proving that water's no problem when money's no object. With oil and cash, Persian Gulf governments can afford to do what most water-scarce nations can't even imagine: desalinate seawater for almost all their freshwater needs.

The WWF also stated that economic wealth does not automatically mean plenty of water. In its report they revealed that some of the world's wealthiest cities—such as Houston or Sydney—are using more water than can be replenished.

According to the *National Geographic,* an estimated 1.2 billion people drink unclean water, and about 2.5 billion lack proper toilets or sewage systems. Additionally, each year more that five million people die from water-related diseases including cholera and dysentery. They state, "All over the globe farmers and municipalities are pumping water out of the ground faster than it can be replenished."[23]

## Urbanization

As the U.S. has undergone a steady process of urbanization, scientists are becoming more concerned about the long-term effects. Unlike rural communities, urban sprawl completely transforms the landscape and the soil and alters the surrounding ecosystem and the climate. Marc Imhoff, a biologist at NASA's Goddard Space Flight Center, and a team of researchers have been looking for ways to measure the effects of urbanization on the biological productivity in the U.S. and other countries around the world. The researchers created a method of mapping urbanization on a countrywide scale by using satellite images of the light cities generate at night. With the resulting city lights maps, they are now zeroing in on the impacts urban sprawl has on the food we eat, the air we breathe and the ecosystem within which we live.

Today, half the world lives in towns and cities. In 2030 the urban population is projected to reach five billion or 60% of the world's population.[24] In U.K. Parliament hearings during May 2006, they pointed out that 100 million people are added to the urban communities of the developing world each year, which is growing three times faster than rural areas—or by 3% a year. As a result, "Nearly one in three city dwellers—almost one billion people—live in slums in conditions characterized by overcrowding, little employment or security of tenure, poor water, sanitation and health services, and widespread insecurity ..."[25]

Moreover, nearly all the population growth in the future is expected to take place in the cities of the less developed world, whose combined population is on course to more than double from 1.9 billion in 2000 to nearly four billion in 2030.[26] Worse, in many of these urban centers, especially in the least developed countries, a high proportion of the residents live in horrible conditions. The UN Habitat report, 'The Challenge of the Slums: Global Report on Human Settlements 2003' states nearly one billion people (or one in three of the world's urban population) live in slums—a figure which may rise to two billion if serious action is not taken. Today, of those one billion people, 600,000 of them cannot even meet their basic needs for shelter, water or health from their own resources.[27]

One city in extremely bad shape is Abidjan in the Ivory Coast. This heavily populated city is the economic capital of the country and home to millions of people. Surrounding the city are a dozen 100-hectare garage dumps. In August of 2006, many of the city's residents started to get sick. The residents uncovered that a Dutch-based company, Trafigura Beheer BV, had supposedly dumped 528,000 liters of liquid chemical waste in the landfills.[28] Government health officials announced that nearly 16,000 people had sought medical treatment and at least six people had died. However, death is nothing new to the people living there. The news article states:

"The main landfill for Abidjan's population of around five million, opened 41 years ago. As Abidjan grew, so did the heaps of trash. Since civil war broke out in Ivory Coast four years ago, the city had ballooned in size to accommodate an untold number of internal refugees. It didn't take long for the effects to be felt in the village. One villager said: 'People die every day here. Old people and children die from things like typhoid and yellow fever.' According to the village's own census, from a population of about 5,000, Akouedo had only one resident over the age of 70, and only 30 had made it to their sixties."[29]

## The world's diseases

Due to overpopulation and the way humans are interacting with other animals in their environment, new infectious diseases are now emerging at an exceptional rate, scientists told a leading conference in St. Louis, Missouri, U.S.[30] According to a report from BBC news, humans are accumulating new pathogens at a rate of one per year. Most of these new infectious diseases, such as avian influenza (Bird Flu) and HIV/AIDS, are coming from other animals. Other interactions occur due to changes in land use through, for example, deforestation which can bring humans into contact with new pathogens. Changes in agriculture can also bring harmful effects, such as the use of exotic livestock. Other important drivers include global travel, global trade and hospitalization.[31]

In Athens, Greece, due to overpopulation which has caused cramped housing conditions and air pollution, the city has given rise to a 'super breed' of mosquito that is larger, faster and more adept at locating human prey. A Yahoo news report states that Athens-based mosquitoes can detect humans at a distance of 25-30 meters and also distinguish colors, unlike their color-blind counterparts elsewhere in the country that only smell blood at 15-20 meters.[32] These 'super mosquitoes' also beat their wings up to 500 times a second—compared to 350 beats for other variations—and are larger by 0.3 micrograms on average, the article said. These mosquitoes of Athens "have adapted to deal with air pollution and insect repellents, and overpopulation in the Greek capital of over four million has provided them with a healthy food supply."[33]

Staying on the topic of mosquitoes, malaria is one of the world's most common and serious tropical diseases. There are at least 300 million acute cases of malaria that occur worldwide each year, resulting in more than one million deaths annually.[34] One study even estimated that 515 million cases of clinical malaria occurred in 2002. Although the majority of the cases occur in the most resource-poor countries, more than half of the world's population is at risk of acquiring malaria. Furthermore, this proportion increases each year because of climate change, worsening health systems, growing drug and insecticide resistance, climate change, natural disasters and armed conflict.[35]

Sub-Saharan Africa, and often pregnant women and children under five years old, are the hardest hit. Malaria accounted for one in ten deaths among children in developing countries in 2002.[36] Approximately 60% of all cases

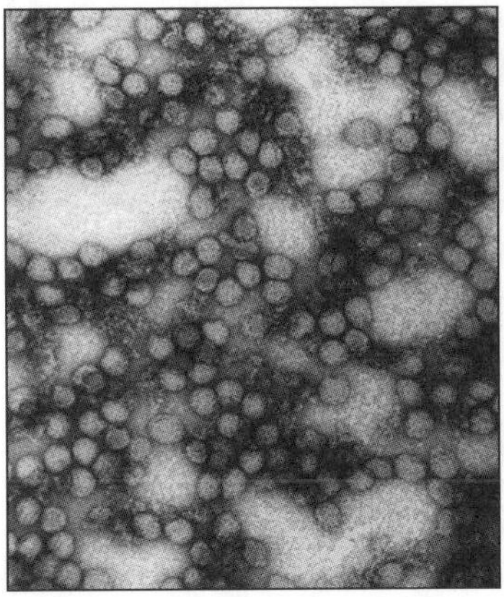

Yellow fever.

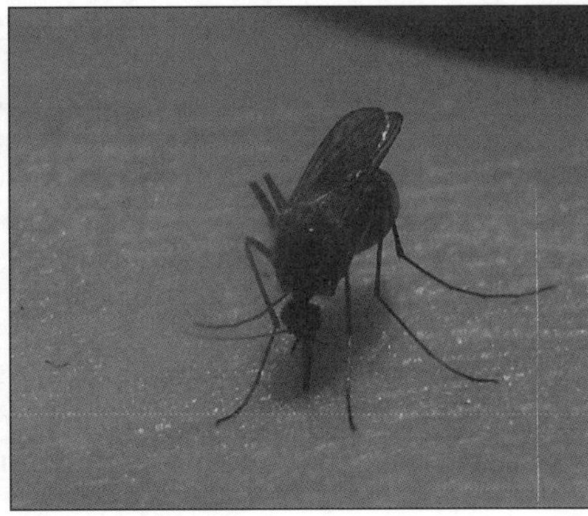

Mosquito, close-up.

of malaria occur among the poorest 20% of the world's population.

Although there are several different treatments for malaria, widespread resistance to conventional anti-malarial drugs has contributed to more causes and deaths. Multi-drug-resistant malaria is now prevalent in many parts of the world, with the highest rate of drug resistance in Southeast Asia.[37] The Global Health Reporting explains that, "The inappropriate use of anti-malarial drugs has contributed to the current situation. Over the past decade, a new group of anti-malarials—the artemisinin compounds—have been used in combination with other drugs and on an increasingly large scale.

"These therapies produce a rapid response and are active against multi-drug-resistant malaria. As a response to increasing levels of anti-malarial resistance, the World Health Organization (WHO) now recommends that all countries experiencing malaria cases that are resistant to conventional therapies use combination therapies."[38]

Quickly switching topics, HIV/AIDS still continues to be a major problem all around the world and it is showing no signs of slowing down. HIV/AIDS is so far out of control that fighting the disease has become the number one priority in Africa. Monetary donors now realize that HIV/AIDS is so enormous that it affects every other issue from education to governance. Due to this severity, almost all foreign aid now goes to the prevention and containment of the disease.

PEPFAR or the President's Emergency Plan for AIDS Relief is the U.S.'s international health initiative addressing the issue. In 2003, the U.S. government approved the $15-billion-dollar package over five years to help fight the disease. As helpful as this sounds, many have criticized the U.S. over their HIV plan. Stephen

Lewis, UN Special Envoy for AIDS, believes that the main problem with the Bush Administration's effort is that it is way too focused on abstinence. Promoting abstinence involves changing human behavior, which can be a very hard thing to do. Mr. Lewis also feels that Washington is telling African nations how to fight HIV/AIDS, which should be left to the individual governments to decide.

Criticisms aside, HIV/AIDS is a huge global problem and I feel it is important to highlight the latest statisitcs.

**Regional/Global HIV/AIDS Statistics**

*Statistics include adults (age 15+) and children, end 2005*

| Region | Living with HIV/AIDS, end 2005 | New HIV infections | Adult (age 15–49) | Deaths due to AIDS in 2005 |
|---|---|---|---|---|
| Sub-Saharan Africa | 24.5 million | 2.7 million | 6.1% | 2.0 million |
| South/ Southeast Asia | 7.6 million | 830, 000 | 0.6% | 560, 000 |
| Latin America | 1.6 million | 140, 000 | 0.5% | 59, 000 |
| East Africa | 680, 000 | 97, 000 | 0.1% | 33, 000 |
| North America | 1.3 million | 43, 000 | 0.8% | 18, 000 |
| Eastern Europe/Central Asia | 1.5 million | 220, 000 | 0.8% | 53, 000 |
| Western/ Central Europe | 720, 000 | 22, 000 | 0.3% | 12, 000 |
| Caribbean | 330, 000 | 37, 000 | 1.6% | 27, 000 |
| North Africa/ Middle East | 440, 000 | 64, 000 | 0.2% | 37, 000 |
| Oceania | 78, 000 | 7, 200 | 0.3% | 3, 400 |
| **Global** | **38.6 million** | **4.1 million** | **1.0%** | **2.8 million** |

(Source: UNAIDS 2006 Report on the global AIDS epidemic, May 2006)

**Antiretroviral coverage**

*Antiretroviral (ARV) coverage in low and middle income countries: Percent on ARVs (of those who need them)* as at December 2005*

| | |
|---|---|
| Sub-Saharan Africa | 17% |
| Latin America and the Caribbean | 68% |
| Europe and Central Asia | 13% |
| North Africa and the Middle East | 5% |
| East, South and Southeast Asia | 16% |
| **All WHO regions**** | **20%** |

* Data are for only developing and transitional countries.
** For a breakdown of WHO regions, please see the List of WHO Member States by Region and Mortality. *Source:* WHO Progress on Global Access to HIV Antiretroviral Therapy: A Report on '3 x 5' and Beyond. March 2006

## Other resource depletion

A coal power station.

According to a study published in the Proceedings of the National Academy of Science, supplies of copper, zinc and other metals have been determined as finite resources. Even if they are recycled, they won't meet the needs of the global population forever.[39]

The researchers used copper stocks in North America as a starting point and tracked the evolution of copper mining, use and loss during the 20th century. They then applied their findings and additional data to an estimate of global demand for copper and other metals if all nations were fully developed and used modern technologies.[40] They concluded that while these metals, such as platinum, are not at risk of depletion in the immediate future, they risk depletion in this century because there is no suitable substitute for use in devices such as catalytic converters and hydrogen fuel cells.[41] For many metals, if the average rate of use per person continues to rise, even the more plentiful metals may face similar depletion risks in the future.

Looking at another resource, the diminishing oil supply is a very large problem. Oil is currently not being discovered at the same rate that it is being used. "For every two barrels pumped out of the ground, the [giant oil companies] find less than one barrel to add to add to the reserves. Production in the United States peaked thirty years ago. Russia peaked in 1987. North Sea production appears to be peaking now. [About] 70% of the oil consumed today was found twenty-five years ago or longer ... Meanwhile, demand for oil keeps moving up [and] the era of cheap and abundant oil is drawing to a close."[42]

As the demand for oil continues to grow, so does the bargaining power of the countries that possess this important resource. Kegley and Wittkopf proved this point by examining the oil reserves. "Saudi Arabia remains the kingpin of oil. It has a quarter of the world's proven reserves, about 262 billion barrels of oil, under its vast desert expanses. Add in the reserves held by the country's allies in the OPEC cartel, and the total is a whopping 815 billion barrels—some three-quarters of world reserves. In contrast, OECD countries, which use much of the world's oil, have less then one-tenth of total reserves."[43] This distribution of oil reserves means that "unless vast reserves are discovered outside the Middle East, global dependence on Persian Gulf oil will grow [and] this is a long-term problem [especially] for the United States in the wake of September 11 because for three decades Americans have only haphazardly tried to fortify themselves against a catastrophic cutoff of oil from the Middle East, which accounts for about a third of world production and two-thirds of known reserves."[44]

Looking closer at the U.S., it has roughly 5% of the world's population, but consumes roughly 26% of the oil. Of this oil, about 64% of it is imported from the Middle East.[45] Renewable energy activist Sean Stalpes writes, "Once oil production peaks, and many energy experts project this will occur by 2030, the prices of fossil fuels will continue to hit unprecedented marks, and we will be left without a sustainable energy infrastructure while much of the rest of the world has long since passed the implementation stages of their growing renewable energy programs."[46]

## Current status of alternative energy

The good news regarding world energy consumption is that it is slowly changing. Little by little people are starting to realize that what has worked in the past 125 years or so, will not work in the near future. This primary energy for this new global energy system will come from some of the most abundant resources on Earth. This includes the sun, the wind and other renewable resources of energy [such as] hydrogen, the lightest and most abundant element in the universe.[47] As you read this, newer technologies are also being worked on. For example, the Japanese Space Agency is currently developing a new form of solar-power generation system, similar to what is commonly used to power artificial satellites. According to the article, a new study, called LE-NET plan, pictures solar panels to be installed in space or the desert, where electricity is generated and is converted into a high-energy laser to be transmitted via laser networks to recipients.[48]

The article states: "Solar-power generation performs more effectively in locations where the climate is less changeable, such as deserts and/or space. Especially in space, where there is no interference by 'night and day' changes, nor by atmosphere, a stable power supply is highly prospective. In this space solar-power concept, a large satellite carrying solar panels will be placed in orbit at an altitude of 36,000 kilometers. Electricity generated by the satellite will be first transmitted in the form of a laser to an airship floating

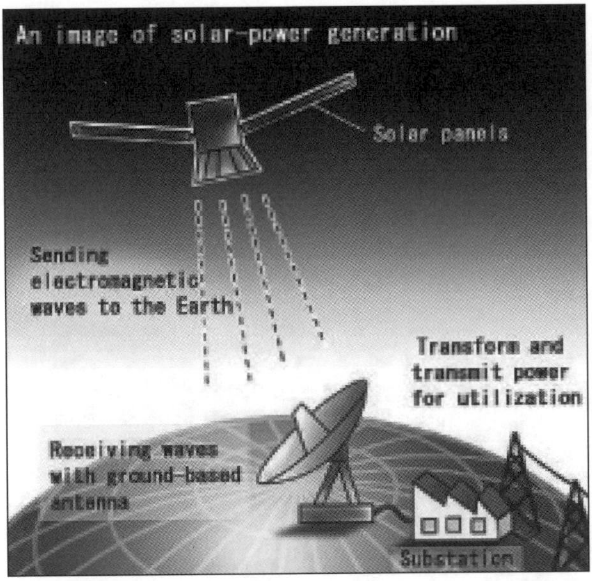


A picture describing how the new solar-power technology would work. The first stage, research is underway with the aim to launch a satellite to verify the technology to transmit electric power from space to the ground using microwaves between 2007 and 2008.

in a lower orbit at 20 kilometers high, and then to ground-based antennas as microwaves or via optical fibers. While the plan still has to deal with many challenges, such as launch costs for solar panel construction and technical studies of laser transmission, the project team currently expects to complete basic laser technologies in the next ten years and has set a goal to conduct its first power transmission test in 2025."[49]

Today, many European countries have started down the right path by beginning to use alternative energy sources.

According to recent article from Bloomberg News, China added almost 500 megawatts of wind energy capacity in 2005, more than double the previous year, as reported by the Global Wind Energy Council. That compares with growth of 11% in Germany, the world's largest wind market, where capacity reached 18,428 megawatts.[50] Nevertheless, other countries are very far behind, especially the U.S. Recently the U.S. has made attempts to initiate alternative energy programs but the two major sources that the Bush Administration advocates are hydrogen and corn-based ethanol. Unfortunately, both sources are not cost-effective solutions. Hydrogen is currently derived from natural gas, an imported fossil fuel, and the technology is not yet good enough to produce it from renewable energy sources. Corn-based ethanol is a good short-term solution, but deriving ethanol from corn is very energy exhausting, requires a lot of cropland, and pollutes water due to fertilizer run off.[51]

On a positive note, hydrogen will most likely be the major competitor in the future energy economy. On the negative side, this will only be a long time from now. Sean Stalpes writes, "Our energy crisis is currently too severe to place the majority of the dollars for renewable energy research and development (R&D) into a source that is decades away from being cost-competitive and environmentally friendly."[52]

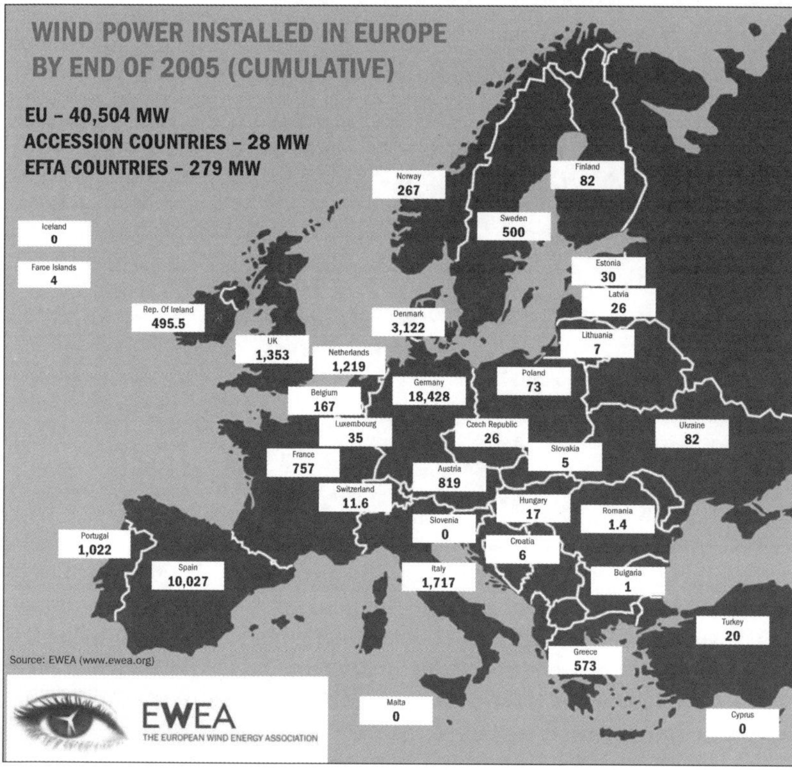


# Is it ethical to have children?

Lucy Siegle, writer for the *Guardian*, asks the question, "Are parents contributing to the future of our planet—or just fuelling an unsustainable population explosion?" This is a very important question and the main reason why overpopulation has become so controversial. Should governments regulate the freedom to have children?

Siegle points out that the issue of how many children to have did not bother Jacob in the book of Genesis. He had 12 sons. It wasn't a problem

Little Grace Elizabeth DiMattia with the author, in March of 2006.

for the famous Von Trapps and the Waltons either. Both of their families had seven children—the same number of kids as the average family in Sudan (one of the poorest countries in the world) attempts to sustain today.[53]

In 1968, human ecologist Garrett Hardin wrote, "what is the necessity of abandoning the commons in breeding. Freedom to breed will bring ruin to all ... The only way we can preserve and nurture other and more precious freedoms is by relinquishing the freedom to breed."[54] On the contrary, others believe that the right to produce children is godgiven and that no one really knows how many people the Earth could hold.

Professor Omer Moav, an economist at the Hebrew University in Jerusalem, feels the answer to combating global poverty lies in having fewer children while, scientists Professor Chris Rapley, director of the British Antarctic Survey and Professor Guillebaud, chair of the Optimum Population Trust, have called for urgent discussions on population management. All three share the opinion that the world cannot sustain an ever-growing global population, even with dramatic lifestyle alterations which might mitigate pressure on life-sustaining resources like shelter, food and water.[55] What do you think?

**Alarming Statistics**[56]

- ❑ Almost 200,000 people are added to the world population every day.
- ❑ One billion new jobs must be created over the next decade just to maintain current employment levels. The availability of a young, educated labor force can be a bonus in newly industrializing countries, but jobs are especially hard to create in countries with high levels of under-employment, poor educational standards and limited infrastructure—and these are often the ones with rapid increases in population.
- ❑ Since 1972, the main driving force leading to pressure on land and water resources has been increasing food production. In 2004, food is needed for some 2.4 billion more people than in 1972.
- ❑ Water requirements will increase as much as 20% by 2025. Approximately one half of the wetlands around the world have been lost since 1900.
- ❑ About five million people die every year from illnesses associated with organic wastes.
- ❑ In the U.S., consumption of materials (wood, metals, synthetics, etc.) has grown 18-fold since 1900.

Although these statistics are horrific, the hardest-hit area is by far the environment. The UN Millennium Development Report from 2005 states "sustainability will not be achieved with current patterns of resource consumption and use. Land is becoming degraded at an alarming rate. Plant and animal species are being lost in record numbers. The climate is changing, bringing with it threats of rising sea levels and worsening droughts and floods. Fisheries and other marine resources are being overexploited."[57] We will now take an in-depth look at the current state of our environment.

# Chapter 3

# The global environment

"We must use time wisely and forever realize that the time is always ripe to do right."
—Nelson Mandela, former president of South Africa

The booming population has led to many problems; however none is as severe as its effect on the environment. Our climate is changing. It has been changing for many decades but it has only been in the last few years, when events like Hurricane Katrina have taken place that people have opened their eyes to see the extent of the damage we have done to the planet. Today it feels as if the future is unfolding right in front of us. Disappearing glaciers, huge storms and fires, and fatal heat waves are clear evidence that the environment has taken a change for the worst. Climatologists reporting for the UN Intergovernmental Panel on Climate Change (IPCC) say we are seeing global warming caused by human activities and there are growing fears of feedbacks that will accelerate this warming.[1]

## Global warming

Global warming is happening. In the beginning many people were denying its existence, but not anymore. While researching for this book I found thousands of studies that all pointed to the same conclusion: global warming is not a lie and that we humans are at least partly to blame.

The problem really started in the late 18th century when the invention of power-driven machinery produced the Industrial Revolution. Since then human-gases, such as the exhaust fumes from our automobiles, have altered the Earth's atmosphere. Commonly known as the 'greenhouse effect', carbon dioxide ($C0_2$), methane ($CH_4$) and other greenhouse gasses form the equivalent of a greenhouse roof by trapping heat emitted from Earth that would otherwise escape into outer space.[2] This causes the Earth's temperature to rise.

According to the European Union (EU), in the 20th century the global average temperature rose by about 0.6°C, and the mean temperature in Europe increased by more than 0.9°C.[3] NASA reports that the five warmest years ever measured were (in decreasing order): 2005, 1998, 2002, 2003, and 2004. In addition, 2006 was the hottest year on record in the U.S. and the sixth-hottest year globally.[4] The worst part is that due to slow reactions of the climate system, past and current greenhouse gas emissions will lead to further temperature increases during the 21st century. If actions are not taken, the EU predicts that global temperatures are expected to increase by 1.4 to 5.8°C above 1990 temperatures by the year 2100, and by 2 to 6.3°C in Europe. "These increases may seem little, but are significant if compared to the average global temperature during the last Ice Age (from about 70,000 to 11,500 years ago): it was only 5-6°C colder than today."[5]

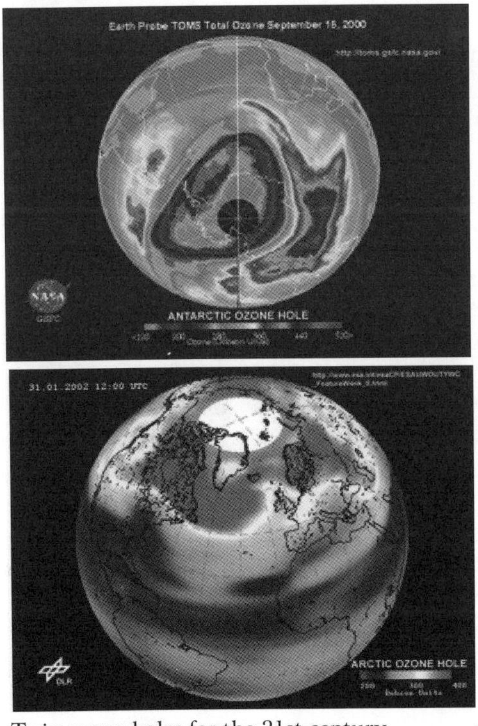

Twin ozone holes for the 21st century.

The rising Earth temperature is slowly melting the ice on both the North and South Pole.

## Deforestation

Another important component of global warming is the world's forests. Trees and plants purify the planet's air of harmful substances like $CO_2$ and hence are key to regulating the climate. Trees are 50% carbon. When they are felled and burned, the $CO_2$ they store escapes back into the air. Deforestation leads to huge releases of $CO_2$ with an estimated one-quarter of the world's $CO_2$ emissions coming from forest clearing.[6] Besides being beneficial to the global atmosphere, they are also vital to the local populations. Forests help control erosion, steady provision of water across rainy and dry seasons, and regulate rainfall.[7]

The World Watch Institute based in Washington D.C. found that global losses of forest area have marched in step with population growth for much of human history. They state, "The two trends rose slowly for millennia, turned upward in recent centuries, and accelerated sharply after 1900. Indeed, 75% of the historical growth in global population and an estimated 75% of the loss in global forested area have occurred in the 20th century. The correlation makes sense, given the additional need for farmland, pastureland, and forest products as human numbers expand. But since 1950, the advent of mass consumption of forest products has quickened the pace of deforestation."[8] Today, FAO estimates that some 13 million hectares of forests worldwide are lost every year, almost entirely in the tropics.[9] Deforestation remains high in Africa, Latin America and Southeast Asia. Another article states, "It is generally agreed that population growth is a primary underlying cause of forest decline, interacting with poverty, corruption, inequitable access to land, and wasteful consumption alongside growing demands for wood products. The dominant force in forest loss is probably the demand for farmland for subsistence farming, but pressure on forestland is also growing under the twin pressures for food and fuel, including soya for animal feed and biofuels as a replacement for oil."[10]

Wood is the main source of energy for nearly three billion people on this planet. Due to facts like these, nearly 200 experts, mostly from developing countries, met in Rome in late August 2006 to address this issue in a workshop organized by the United Nations Framework Convention on Climate Change (UNFCCC) and hosted by FAO. "We are working to solve two of the key environmental issues—deforestation and global warming—at the same time," said FAO Senior Forestry Officer Dieter Schoene.[11]

The delegates showed their readiness to act on deforestation, 80% of which is due to increased farmland to feed growing populations.[12] They felt that part of the solution is to increase agricultural productivity so that there is less demand to convert forests into farmland. Moreover, the delegates stressed that they needed financial help from the developed world.

## Desertification

Orphaned ship in former Aral Sea near Aral, Kazakhstan.

In Chapter 2, we briefly touched on the problem of the loss of land to grow food. Most of this loss is due to desertification. Kegley and Wittkopf define desertification as, "the creation of deserts due to soil erosion, overfarming and deforestation, which converts cropland to non-productive, arid sand."[13]

If the land's topsoil is mistreated, it can be blown and washed away in a few seasons. For it to regenerate takes centuries. According to the UN, this gradual process of soil productivity loss and the thinning out of the vegetative cover is mainly due to human activities and climatic variations such as prolonged droughts and floods. In addition, overexploitation is generally caused by economic and social pressure, ignorance, war, and drought.[14]

Although you would think desertification is confined to Africa, where two-thirds of the continent is desert or drylands, it is not. Desertification is a global problem. It directly affects 250 million people and a third of the earth's land surface or over four billion hectares.[15] Additionally, the livelihoods of some one billion people who depend on land for most of their needs, who usually are the world's poorest in over one hundred countries, are very much threatened. In the U.S., over 30% of the land is affected by desertification.[16] Furthermore, as the UN points out, "One quarter of Latin America and the Caribbean is deserts and drylands. In Spain, one fifth of the land is at risk of turning into deserts. The growing severity of the threat in the Northern Hemisphere is also illustrated by severe droughts in the United States and water scarcity in southern Europe. In China, since the

1950s, sand drifts and expanding deserts have taken a toll of nearly 700,000 hectares of cultivated land, 2.35 million hectares of rangeland, 6.4 million hectares of forests, woodlands and shrub lands. Worldwide, some 70% of the 5.2 billion hectares drylands used for agriculture are already degraded and threatened by desertification."[17]

In these areas where desertification occurs, the results are devastating. The loss of land may force farmers to seek a way of living in more fertile lands or cities. Some 135 million people are at risk of being displaced as a consequence of desertification, which includes the 60 million people who are expected to eventually move from the desertified areas in Sub-Saharan Africa towards northern Africa and Europe in the next 20 years.[18] In North America, between 700,000 to 900,000 Mexicans leave their rural dryland homes to find a living as migrant workers in the U.S. every year. Also, desertification can play a large role in social and political conflicts. Half of the 50 armed conflicts in 1994 had environmental causal factors characteristic of the drylands.[19]

Desertification also leads to other environmental problems. This includes: making land areas flood-prone; causing increased soil salinity; the deterioration of the quality of water; silting of rivers, streams and reservoirs; and the drying up of rivers that feed large lakes (the Aral Sea and Lake Chad have both seen their shorelines shrink dramatically in this way).[20] Moreover, both desertification and deforestation strongly affect the world's plant and animal species. Forests provide habitat to a diverse selection of wildlife. Tropical forests alone are home to more than 50% of the world's plant and animal species.[21]

## Extinction of species

Scimitar-horned oryx (*Oryx dammah*) were once one of the most common large mammals of northern Africa. Overhunting for the animal's meat, hide and magnificent horns, combined with habitat loss caused major declines in the species. By the end of the 20th century none were known to remain in the wild. Currently listed as Extinct in the Wild, the species is now part of a major captive breeding and reintroduction program.

Every year the World Conservation Union (IUCN) publishes its Red List of Threatened Species. The 2006 list revealed the ongoing decline of the status of plants and animals. The report states, "The total number of species declared officially *Extinct* is 784 and a further 65 are only found in captivity or cultivation. Of the 40,177 species assessed using the IUCN Red List criteria, 16,119 are now listed as threatened with extinction. This includes one in three amphibians and a quarter of the world's coniferous trees, on top of the one in eight birds and one in four mammals known to be in jeopardy."[22]

The IUCN Red List threat categories are the following, in descending order of threat:

- *Extinct or Extinct in the Wild;*
- *Critically Endangered, Endangered and Vulnerable*: species threatened with global extinction;
- *Near Threatened*: species close to the threatened thresholds or that would be threatened without ongoing specific conservation measures;
- *Least Concern*: species evaluated with a low risk of extinction;
- *Data Deficient*: no evaluation because of insufficient data.

(Source: IUCN Webpage: http://www.iucn.org)

Polar bears, the world's largest bears, are facing a bleak future because of global warming.

One animal we are familiar with is the polar bear. Unfortunately polar bears are set to become one of the most notable casualties of global warming. The IUCN reports that the impact of climate change is increasingly felt in polar regions, where summer sea ice is expected to decrease by 50-100% over the next 50-100 years.[23] The polar bears are highly dependent upon Arctic ice-floes for hunting seals and are highly specialized for life in the Arctic marine environment. Due to this they are predicted to suffer more than a 30% population decline in the next 45 years.[24] Previously listed by IUCN as a *conservation dependent* species, the polar bear moves into the threatened categories and has been classified as *Vulnerable*.

Another animal that most are familiar with is the large freshwater species the hippopotamus or the 'hippo'. Sadly enough they also face difficulty. One of Africa's best-known aquatic icons, the hippo has been listed as threatened for the first time and is now classified as *Vulnerable*, primarily because of a catastrophic decline in the Democratic Republic of the Congo (DRC).[25] The IUCN states, "In 1994 the DRC had the second-largest population in Africa—30,000 after Zambia's 40,000—but numbers have plummeted by 95%. The decline is due to unregulated hunting for meat and the ivory of their teeth."[26]

According to a report from 2005 entitled, 'Millennium Ecosystem Assessment' (MA), organisms are disappearing at around 100 to 1,000 times the "background levels" seen in the fossil record.[27] Dr. Georgina Mace, the Director of Science at the Institute of Zoology, in London, U.K., an MA synthesis team member explained, "Changes in biodiversity were more rapid in the last 50 years than at any time in human history." Mace added, "And when you look to the future, to various projections and scenarios, we expect those changes to continue and in some circumstances to accelerate ... Future models are very uncertain but all of them tell us that as we move into the next 100 years, we'll be seeing extinction rates that are a thousand to 10,000 times those in the fossil record."[28]

Scientists have also warned that removing so many species puts our own existence at risk. On a positive note, "Most of the approaches to achieving more sympathetic management of the natural environment and the conservation of biodiversity—I think we and governments know them already," commented Graham Wynne, the chief executive of a U.K. bird conservation group. He added, "And you can't get away from the fact that we simply need more money ... The sums of money we throw at the environment in the West are relatively modest; and the sums of money the West is prepared to devote to developing countries is pitiful."[29]

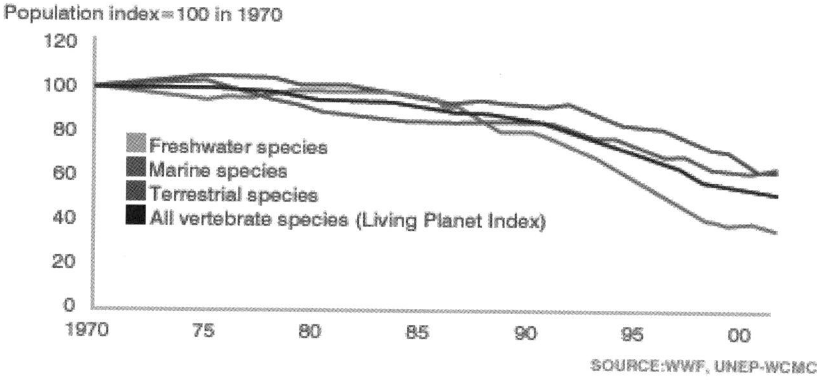


The Living Planet Index is a measure of the state of the world's biodiversity. It measures trends of vertebrate populations in terrestrial, freshwater and marine environments.

If nothing is done in the future, global warming in the 21st century could mean a return of temperatures last seen in the age of the dinosaur. This could lead to the extinction of up to half of all species. Chris Thomas of the University of York explains that not only will $CO_2$ levels be at their highest levels in 24 million years, but global average temperatures will be higher than they have been for up to ten million years.[30] Additionally, between 10 and 99% of species will be faced with atmospheric conditions that last existed before they evolved, and as a result from 10-50% of them could disappear. "We may very well already be on the breaking edge of a wave of mass extinctions," Thomas told the annual meeting of the British Association for the Advancement of Science.

If the global temperatures rise by 6°C by 2100 (extreme case) due to global warming, "we will be going back to global temperatures not seen since the age of the dinosaur," Thomas said. "We are starting to put these things into a historical perspective. These are conditions not seen for millions of years, so none of the species will have been subjected to them before" he added.[31] Scientific observations have found that 80% of species have already begun moving their traditional territorial ranges in response to the changing climatic conditions.[32] Some examples from the UNFCCC are as follows:

- ❑ Scientists have observed climate-induced changes in at least 420 physical processes and biological species or communities.
- ❑ In the Alps, some plant species have been migrating upward by one to four meters per decade, and some plants previously found only on mountaintops have disappeared.
- ❑ In Europe, mating and egg-laying of some bird species has occurred earlier in the season—in the U.K., for example, egg-laying by 20 of 65 species, including long-distance migrants, advanced by an average of eight days between 1971 and 1995.
- ❑ Across Europe, the growing season in controlled, mixed-species gardens lengthened by 10.8 days from 1959 to 1993. Butterflies, dragonflies, moths, beetles, and other insects are now living at higher latitudes and altitudes, where previously it was too cold to survive.[33]

Thomas also notes that climate-triggered fungal pathogen outbreaks had already led to the extinction of more than 1% of the planet's amphibian species. In addition, "Not only would some species simply find no suitable space to live anymore, but there would be confrontations with invasive species being forced to move their territory. This would produce not just wipe-outs but species' mixtures never seen before. And the changes would all happen at a faster rate than ever before in evolution … In geological terms 100 years is effectively instantaneous."[34]

## Our oceans

Due to climate change, the Pacific and Atlantic oceans have been warming since the 1950s; the Indian Ocean has warmed since the 1960s.[36] The UNFCCC informs us of some startling facts from the 20th century:

- ❑ Arctic air temperatures increased by about 5°C during the 20th century—ten times faster than the mean global surface temperature. In the Russian Arctic, buildings are collapsing because permafrost under their foundations has melted.
- ❑ Snow cover has declined by some 10% in the mid- and high latitudes of the Northern Hemisphere since the late 1960s. The annual duration of lake and river ice cover apparently shortened by about two weeks during the 20th century.
- ❑ Almost all mountain glaciers in non-polar regions retreated during the

20th century. The overall volume of glaciers in Switzerland decreased by two-thirds.[37]

"The most visible and familiar impact of human activity on the marine environment is oil pollution caused by tanker accidents. Yet despite the scale and visibility of such impacts, the total quantities of pollutants entering the sea from oil spills are dwarfed by those of pollutants introduced from other sources. These include domestic sewage, industrial discharges, urban and industrial run-off, accidents, spillage, explosions, sea dumping operations, mining, agricultural nutrients and pesticides, waste heat sources, and radioactive discharges."[35]

This rise in both water and air temperatures has had other implications. Today Pacific, North Atlantic and Indian Ocean oscillations are being tracked by scientists worldwide. Scientists further track the surface waves sloshing through the Pacific near the equator.[38] One consequence of these oscillations is a rise in infectious diseases. Each infectious disease can now extend to places they never had a chance to travel to before. Because of this, an increase in incidence rates of infectious diseases in the last three and half decades has been observed.[39] Sinyan Shen states, "The propagation pattern of these agents is highly affected by temperature and humidity. Many tropical and subtropical diseases for example are not commonly present in regions where extended cold weather renders these agents inactive. This is no longer true when we have extreme climatic swings."[40]

Rising ocean temperatures, combined with melting ice on land, is also raising sea levels. Today these higher ocean levels are already contaminating underground water sources in Israel and Thailand, in various small island states in the Pacific and Indian oceans and the Caribbean Sea, and in some of the world's most productive deltas, such as China's Yangtze Delta and Vietnam's Mekong Delta.[41] In the 21st century, human activity could trigger an irreversible melting of the Antarctic glaciers and the Greenland ice sheet. If this occurs, the sea level could rise up to six meters which would

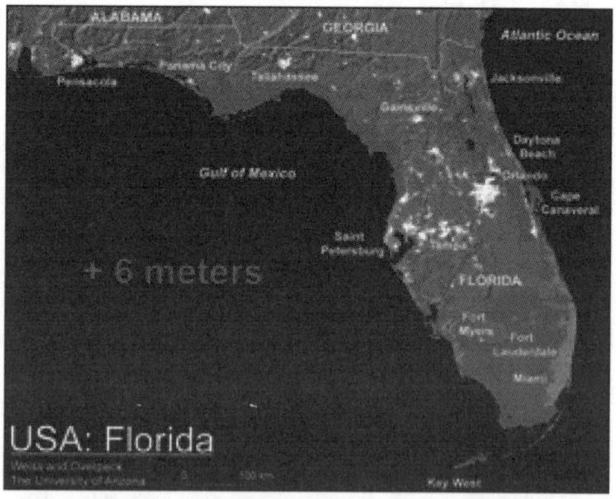


Effect of six meter sea-level rise on Florida, (USA). White areas are population centers: they are dark gray where flooding overlies them.

be enough to cover land occupied by billions of people.[42]

Another environmental problem with our oceans is caused by industrial fishing. Today we have giant ships with computer equipment that can locate huge schools of fish quickly and accurately. These ships have surpassed the ocean's ecological limits.[43] The organization Greenpeace states, "As larger fish are wiped out, the next smaller fish species are targeted and so on. (Canadian Fisheries expert Dr. Daniel Pauly warns that if this continues our children will be eating jellyfish.)"[44]

The problem doesn't only lie with just over-catching. Modern fishing practices are very wasteful as fishing nets kill up to 300,000 whales, dolphins and porpoises globally every year. Greenpeace writes, "Entanglement is the greatest threat to the survival of many species. Moreover, some fishing practices destroy habitat as well as inhabitants. Bottom trawling, for example, destroys entire ancient deep-sea coral forests and other delicate ecosystems. In some areas it is the equivalent of plowing a field several times a year."[45] It is believed that 90% of the large predatory fish in the oceans have gone since the beginning of industrial fishing.[46]

### Extra-strength weather

Residents are evacuated from New Orleans, Lousiana (USA) on August 31, 2005 after Hurricane Katrina.

The warmer air and water temperatures have also increased weather-related natural catastrophes. A survey conducted by the world's largest re-insurer Munich Re shows that between 1994 and 2003 there were almost three times as many weather-related natural catastrophes as in the 1960s. Even without counting the tsunami in South Asia, 2004 was the costliest year for natural catastrophes so far in insurance history, with global insured losses of over US$30 billion.[47]

Current computer models predict that the weather will worsen. Higher temperatures bring with it more evaporation which means a warmer atmosphere. A warmer atmosphere can hold more moisture, and when this occurs, there is more water in the air that has the ability to fall as precipitation, leading to floods and stronger storms. On the other hand, the drier regions can be worse off because if their areas get hotter, their atmosphere is more than likely to lose more moisture. This causes desertification and drought.

The UNFCCC provides two examples to prove these important points. First, in Africa's large catchment basins of Niger, Lake Chad, and Senegal, total available water has decreased by 40 to 60 %, and desertification has increased lower average annual rainfall, runoff, and soil moisture, especially in southern, northern, and western Africa. Secondly, the Rhine floods of 1996 and 1997, the Chinese floods of 1998, the East European floods of 1998 and 2002, the Mozambican and European floods of 2000, and the monsoon-based flooding of 2004 in Bangladesh (which left 60% of the country under water), are an indication that storms indeed are growing more powerful.[48]

**Global emissions in 2000**

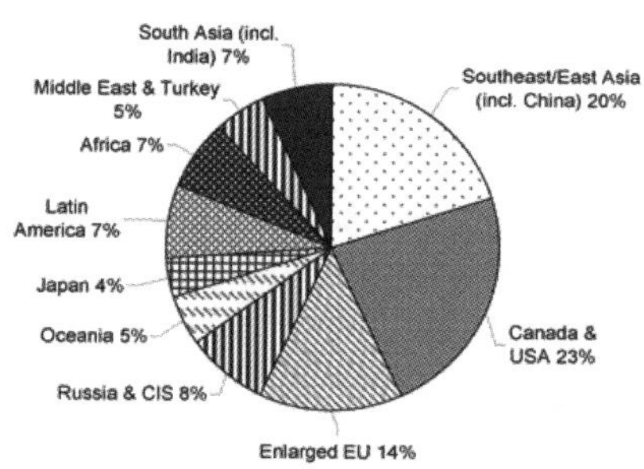


**Global emissions in 2050**

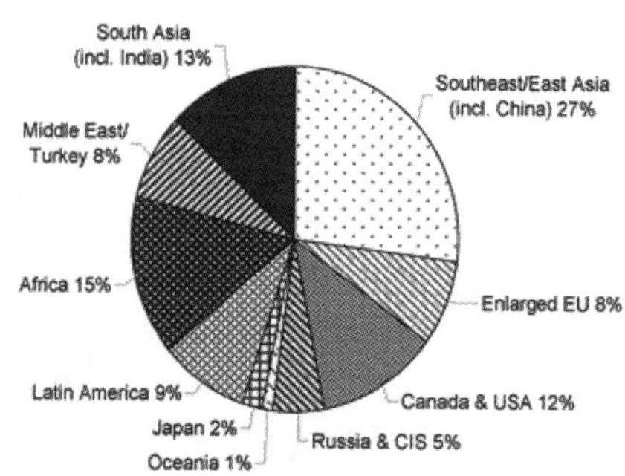


(Source: http://jpn.cec.eu.int)

## Today's actions

Governments around the world are very aware of what is happening to our planet. An Earth Summit was held in 1992. At this conference, the world agreed to prevent 'dangerous' climate change. The first step of this process came in 1997 with the Kyoto Protocol (see Chapter 7 for more information on international treaties). On February 16, 2005, the Kyoto Protocol officially entered into force bringing with it modest emission reductions from industrialized countries.[49] It requires 35 industrialized countries to reduce those emissions by 5% below 1990 levels by 2012.[50]

Although Kyoto is a positive achievement in combating climate change, a UN report from October 2006 indicated that many countries are struggling to meet their targets.[51] Additionally, people understand that more has to be done. For example, many observers say deeper cuts are needed and developing nations, which have large and growing populations, will one day have to join in.[52]

In November of 2006, thousands of delegates from around the world attended the UN conference on climate change in Nairobi, Kenya. Many

Pollution in Kuala Lumpur. Many Asian countries are struggling to deal with the problem of pollution. For example, on November 5, 2005, Beijing's pollution index hit its highest measurement level for the second day as authorities warned people to stay indoors as a murky grey haze settled over the Chinese capital.

at the two-week session were looking for signs that the U.S. might ease its stand against mandatory reductions in emissions that scientists blame for global warming, however, few expect to see such a change while the Bush administration is in power.[53] His administration still objects to Kyoto-style mandates because, it says, they would hamstring U.S. economic growth and because poorer countries are exempted from the controls.

Grace Akumu, head of the advocacy group Climate Network Africa states, "The impacts are increasing every year in frequency and intensity, and every year we're feeling we're less able to cope."[54] Now what is the answer to the problem? Perhaps author Fred Pearce has the right idea when he said, "The world needs to quickly improve the efficiency of its energy usage and develop renewable non-carbon fuels like wind, solar, tidal, wave and perhaps nuclear power. It also means developing new methods of converting this clean energy into motive power, like hydrogen fuel cells for cars."[55]

At this moment in time some things are blatantly clear. In the future, global warming and its effects will be unfair. The UNFCCC states: "The industrialized countries of North America and Western Europe, along with a few other states, such as Japan, are responsible for the vast bulk of past and current greenhouse gas emissions. These emissions are a debt unwittingly incurred for the high standards of living enjoyed by a minority of the world's population. Yet those who will suffer most from climate change will be in the developing world. They have fewer resources for coping with storms, with floods, with droughts, with disease outbreaks, and with disruptions to food and water supplies. They are eager for economic development themselves, but may find that this already difficult process has become more difficult because of climate change. The poorer nations of the world have done almost nothing to cause global warming yet are most exposed to its effects."[56]

In order to stabilize the atmosphere's $CO_2$ levels, which in turn will stabilize temperatures, we will have to start cutting emissions by 70 to 80%. As Fred Pearce stated, "The quicker we do that, the less unbearably hot our future world will be."[57]

# Chapter 4

# Protecting basic human rights

"Education is the most powerful weapon which you can use to change the world."
—Nelson Mandela

Mauthausen Concentration Camp liberated during World War II.

In 1948 the UN issued the Universal Declaration of Human Rights. This declaration set forth human right standards for all nations. It states that: "All human beings are born free and equal in dignity and rights ... Everyone has the right to life, liberty, and security of person."[1] The actual declaration lists more rights than this and, in 1975, the UN added the rights of freedom of movement and freedom to publish and exchange information. The declaration is not a binding agreement, but it showed that people around the world are committed to ensuring that basic human rights are respected.

As great of an achievement as the Universal Declaration of Human Rights was, protecting basic human rights is still a major battle. On every continent, human rights violations continue to occur. Let's begin by focusing on some of these violations.

## Human trafficking

Trafficking human beings is a very large issue and over the past decade it has spiraled out of control. The people who organize and carry out the trafficking (traffickers) face few risks and can earn large profits by taking advantage of large numbers of potential immigrants. The industry is now worth over billions of dollars each year. For example, in Asia, girls from villages in Nepal and Bangladesh—the majority of whom are under 18—are sold to brothels in India for US$1,000.[2]

According to the UN Office on Drugs and Crime (UNODC), "From Himalayan villages to Eastern European cities, people—especially women and girls—are attracted by the prospect of a well-paid job as a domestic servant, waitress or factory worker. Traffickers recruit victims through fake advertisements, mail-order bride catalogues and casual acquaintances."[3] They are even known to lie to parents by telling them their children will have a better life and will be looked after personally by them.

Once they arrive at their destination, the traffickers control the conditions they are in, and the children become dependent on them for food, shelter and other basic necessities. Traffickers also play mind games with the victims by telling them such things like they will inform the government, then they will deport them or put them in jail. They even go as far as threatening the person's family if they do not cooperate.[4] Using these techniques they force the women and even children to work as prostitutes.

The U.S. State Department provides a typical example: "Nepal/India: Reena was brought to India from Nepal by her maternal aunt, who forced the 12-year-old girl into a New Delhi brothel shortly after arrival. The brothel owner made her have sex with many clients each day. Reena could not leave because she did not speak Hindi and had no one to whom she could turn. She frequently saw police officers collect money from the brothel owners for every new girl brought in. The brothel owner coached Reena and all the girls to tell anyone who asked that they were 25 years old and had voluntarily joined the brothel. Reena escaped after two years and now devotes her life to helping other trafficking victims escape."[5]

Trafficking of children violates their human right as a child to grow up in a protective environment free from all forms of abuse and exploitation. These victims often suffer horrible conditions that result in physical, sexual, and psychological trauma.[6] Sexually transmitted diseases and HIV/AIDS are common in the victims being used in prostitution. The State Department writes, "Anxiety, insomnia, depression, and post-traumatic stress disorder are common psychological manifestations among trafficked victims.

Unsanitary and crowded living conditions, coupled with poor nutrition, foster a host of adverse health conditions such as scabies, tuberculosis, and other communicable diseases."[7]

Many believe that human trafficking only deals with the sex industry. On the contrary, people are trafficked to work in sweatshops as bonded labor and men work illegally in the 'three D-jobs'—dirty, difficult and dangerous.[8] However, the sex industry is the hardest hit. A recent U.S. Central Intelligence Agency (CIA) report estimated that between 45,000 to 50,000 women and children are brought to the U.S. every year under false pretenses and are forced to work as prostitutes, abused laborers or servants. The UN's organization dealing with children, UNICEF, estimates that more than 200,000 children are enslaved by cross-border smuggling in West and Central Africa.[9] The actual numbers vary significantly from organization to organization, but what we do know is that the numbers are high.

In 2004, the U.S. State Department became the first to disaggregate transnational trafficking in persons by age and gender using U.S. government data. The data showed that, of the estimated 600,000 to 800,000 men, women, and children trafficked across international borders each year, approximately 80% are women and girls, and up to 50% are minors.[10] However they found that these numbers do not include the millions of victims around the world who are trafficked within their own national borders. In their 2006 report they examined the trafficking of people for purposes of slave labor, often in their own countries.[11] The State Department found that this form of human trafficking is harder to identify and estimate than sex trafficking, yet it may be much greater in size when we count domestic trafficking.

We know now that whether it is transnational or domestic trafficking, it is almost always a form of organized crime. Unfortunately victims of human trafficking have become another article of trade in a larger realm of criminal businesses involving other commodities, such as drugs, weapons and money laundering that generate money or seek to reduce risks for traffickers. The UNODC states, "Trafficked persons should also be seen as victims of crime. Support and protection of victims is a humanitarian objective and an important means of ensuring that victims are willing and able to assist in criminal cases."[12]

Here is another example from the State Department: "Uganda: Michael was 15 when he was kidnapped by the Lord's Resistance Army (LRA) to serve as a combatant in the Ugandan insurgent force. During his forced service in the LRA, he was made to kill a boy who had tried to escape. He also watched another boy being hacked to death because he did not alert the guards when his friend successfully escaped."[13]

Human trafficking not only has a huge impact on the individual, but on other areas as well. It can affect wages, diminished workforce productivity, a loss of remittances, and an undereducated generation.[14] Furthermore, by forcing children to work, you are denying them access to education.

## Lack of Education

Most child victims of human trafficking are denied their right to education, but in total, more than 800 million people are illiterate worldwide. In addition, one hundred million children are not in school, which means they're not learning to read. Furthermore, of the 771 million adults who cannot read a simple book, more than two-thirds of them are women.[15]

On the top of the human rights list is ensuring education for all children. The reasons for this go far back hundreds of years. In 1676, the governor of colonial Virginia complained about how difficult it was to govern people who were "poor, endebted, discontented and armed."[16] In response to this, free public education was established to make new generations economically self-supporting. Today, there are still concerns about religious boarding schools in Pakistan and elsewhere, and to post-9/11 fears of what might be taught there.[17] The Government of Pakistan neither provides education nor does it know who is educating its youth. Just as 350 years ago, public authorities today need to take charge of education because it is simply too dangerous not to do so.

U.S. First Lady Laura Bush said, "By investing in literacy and education, governments build their economies. When people read, they're more likely to participate in business and trade, which leads to greater economic development. And literacy helps both men and women provide for themselves, their families, and their communities. Countries prosper with an educated workforce."[18]

Literacy and education also affects the public health sector. In South Africa, education is saving lives. There the Mothers-to-Mothers-to-Be Program provides HIV-infected women with information and support to prevent their unborn children from contracting the disease. Additionally, those women who have successfully delivered HIV-negative babies serve as mentors to other mothers.[19] Laura Bush adds, "People who can read can understand the label on a food container. They know how to follow instructions on the bottle of medicine. They're also able to make informed decisions about their lives that will keep them and their families healthy."[20]

As helpful as these programs are, the larger issue still remains. Unfortunately even the main actor addressing this issue, UNESCO, has been calling for "greater harmonization and alignment in the approaches of multilateral organizations."[21] Again, we must not forget that education is a global problem. An editorial in *Le Monde* states, "irresponsible policies which have not prepared the youth for the future but have rather loaded a 100 billion euros public debt onto their shoulders." Plus, "the price of this sin of omission is widespread youth unemployment, dotted with occasional outbursts of violence, especially where the fault line between the rich and the poor corresponds with the boundaries of belonging marked by race, religion, language, or provenance."[22] This shows us that even countries like France can have problems addressing the education issue.

Right now education for the youth is guaranteed in the Netherlands and Norway but denied in Cuba and China.[23] The following table shows us how bad the situation is around the globe.

**School-leaving age below the minimum age of employment**
*Officially defined ages at which children start and finish schooling by country*

**Latin America and Caribbean:**
Bolivia 6-13
Dominican Republic 5-13
Haiti 6-11
Honduras 7-12
Jamaica 6-11
Panama 6-11
Suriname 6-11
Trinidad & Tobago 5-11

**Asia and Pacific:**
Afghanistan 7-12
Burma/Myanmar 5-9
Bangladesh 6-10
Laos 6-10
Pakistan 5-9
Philippines 6-12

**Middle East:**
Egypt 6-13
Iran 6-10
Kuwait 6-13

Saudi Arabia 6-11

**Eastern Europe and Central Asia:**
Albania 6-13

**Sub-Saharan Africa:**
Angola 6-9
Benin 6-11
Burundi 7-12
Cameroon 6-11
Cape Verde 6-12
Chad 6-11
Congo/Kinshasa 6-13
Equatorial Guinea 7-11
Eritrea 7-13
Ethiopia 7-12
Guinea-Bissau 7-12
Kenya 7-13
Lesotho 6-12
Malawi 6-13
Mozambique 6-12
Niger 7-12
Nigeria 6-11
Rwanda 7-12
Senegal 7-12
Sudan 6-13
Swaziland 6-12
Tanzania 7-13
Zambia 7-13
Zimbabwe 6-12

*Note*: This table lists contries in which compulsory eduction ends before the global minimum age for employment of 14. The information is authoratative since it was officially forwarded to UNESCO by the respective ministries of education.

(Source: Global Education Digest 2005: Comparing education statistics across the world, UNESCO Institute for Statistics, Montreal, 2005, available at http://www.uis.unesco.org)

## Corruption

One of the main problems with educating the youth, among other things, is corruption. The most commonly thought-of form of corruption is political corruption. When people elect politicians and political parties to office in their government, they expect them to act in the public interest. Once they are in office, they have access to public resources and the power to make decisions that impact our lives. Because they have this power, politicians can inflict major damage by simply acting out of greed, or by paying off the people who helped them get to their current political position.[24] Transparency International (TI) states, "It is not surprising that people all over the world are demanding absolute probity of their political leaders: citizens in three out of four countries polled by TI and Gallup International in 2003 and in 2004 singled out political parties as the institution they perceived as most corrupt."[25] Furthermore, a 2004 TI study acknowledged the fact that there is worldwide concern that money that is supposed to be used for development cooperation is being wasted through corruption.

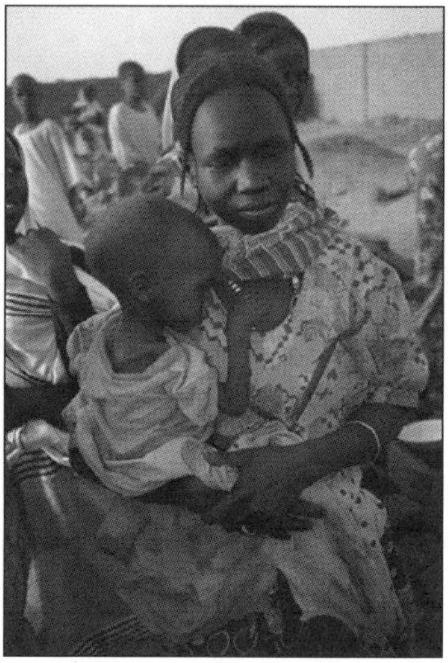
Corruption has taken its heaviest toll in Africa.

Another major part of political corruption is what TI calls the 'revolving door' between business and politics. In running a government, politicians need to contract people publicly or businesses to do certain tasks. Due to this need, it is an enormous area of business. TI states, "Think of pharmaceutical companies vying to supply a government vaccination program, the privatization of a government-owned telecommunications company, or the awarding of contracts to reconstruct destroyed infrastructure in Iraq."[26] The main areas that are especially prone to corruption are the arms industry, construction and engineering, oil, and development aid.

Corruption in public contracting has many consequences including the distortion of fair competition, the waste of scarce resources, and the neglect of basic needs, perpetuating poverty.[27] TI reports that if public contracting corruption is not contained, it will grow. They write, "It is estimated that systemic corruption can add 20-25% to the costs of government procurement, and frequently results in inferior-quality goods and services and unnecessary purchases."[28]

There are many other forms of corruption as well. In most developing countries, low salaries for civil servants is a driving force in why some of them commit corrupt activities. For example, in South Africa police officers are underpaid and because of it, they have been known to perform corrupt tasks. One person I know was threatened by a police officer during a routine traffic stop. The officer told the gentleman that if he didn't go to the ATM right now and withdraw 1,000 rand for him that he would frame him by placing drugs in his car and then throw him in jail. Incidents like this give the police service a bad name. This is completely unfair to the police officers who do work by the book and respect the law, some of whom are my good friends. Sadly, bribes and other gifts and favors may form a significant percentage of a public employee's total income in societies where civil service pay is low.[29]

Corruption can also be present in hiring and job advancement of employees. For instance, someone receiving a job may be determined more by connections and payoffs than by merit, reducing the professionalism and competence of the bureaucracy and strengthening the cycle of corruption.[30]

To help measure corruption around the world, TI heads up two of the biggest studies on the topic. The first study is called the Bribe Payers Index (BPI). On October 4, 2006, TI released the results of the responses of more than 11,000 business people in 125 countries polled in the World Economic Forum's Executive Opinion Survey 2006.

A score of 10 indicates a perception of no corruption, while zero means corruption is seen as rampant.[31] TI found that the wealthiest countries generally rank in the top half of the index, but still routinely pay bribes, particularly in developing economies. The emerging export powers like China, India, and Russia rank among the worst. China, for example, has made efforts to strengthen domestic anti-corruption activities but it has failed to extend abroad. TI Chair Huuette states, "Bribing companies are actively undermining the best efforts of governments in developing nations to improve governance, and thereby driving the vicious cycle of poverty."[32]

The second study, and more well known, is the Corruption Perceptions Index (CPI). The 2006 CPI released in November of 2006 was the largest to date, analyzing 163 countries. The study used multiple expert opinion surveys that polled perceptions of the public sector.[33] The 2006 CPI came to almost the same conclusion as the BPI as the CPI results showed a strong correlation between corruption and poverty. To be more specific, "Almost three-quarters of the countries in the CPI score below five (including all low-income countries and all but two African states) indicating that most countries in the world face serious perceived levels of domestic corruption. Seventy-one countries—nearly half—score below three, indicating that corruption is perceived as rampant. Haiti has the lowest score at 1.8; Guinea, Iraq and Myanmar share the penultimate slot, each with a score of 1.9. Finland, Iceland and New Zealand share the top score of 9.6."[34]

Countries with a significant worsening in perceived levels of corruption include: Brazil, Cuba, Israel, Jordan, Laos, Seychelles, Trinidad and Tobago, Tunisia and the United States. Countries with a significant improvement in perceived levels of corruption include: Algeria, Czech Republic, India, Japan, Latvia, Lebanon, Mauritius, Paraguay, Slovenia, Turkey, Turkmenistan and Uruguay.

**Full Results of the BPI 2006**

| | Rank country/ territory | Number of respondents | Average score (scale 0–10) | Standard deviation | Margin of error (95% confidence) |
|---|---|---|---|---|---|
| 1 | Switzerland | 1744 | 7.81 | 2.65 | 0.12 |
| 2 | Sweden | 1451 | 7.62 | 2.66 | 0.14 |
| 3 | Australia | 1447 | 7.59 | 2.62 | 0.14 |
| 4 | Austria | 1560 | 7.5 | 2.6 | 0.13 |
| 5 | Canada | 1870 | 7.46 | 2.7 | 0.12 |
| 6 | U.K. | 3442 | 7.39 | 2.67 | 0.09 |
| 7 | Germany | 2873 | 7.34 | 2.74 | 0.09 |
| 8 | Netherlands | 1821 | 7.28 | 2.69 | 0.12 |
| 9 | Belgium | 1329 | 7.22 | 2.7 | 0.15 |
| 10 | U.S. | 5401 | 7.22 | 2.77 | 0.07 |
| 11 | Japan | 3279 | 7.1 | 2.87 | 0.1 |
| 12 | Singapore | 1297 | 6.78 | 3.04 | 0.17 |
| 13 | Spain | 2111 | 6.63 | 2.73 | 0.12 |
| 14 | UAE | 1928 | 6.62 | 3.09 | 0.14 |
| 15 | France | 3085 | 6.5 | 3 | 0.11 |
| 16 | Portugal | 973 | 6.47 | 2.79 | 0.18 |
| 17 | Mexico | 1765 | 6.45 | 3.17 | 0.15 |
| 18 | Hong Kong | 1556 | 6.01 | 3.13 | 0.16 |
| 19 | Israel | 1482 | 6.01 | 3.14 | 0.16 |
| 20 | Italy | 2525 | 5.94 | 2.99 | 0.12 |
| 21 | South Korea | 1930 | 5.83 | 2.93 | 0.13 |
| 22 | Saudi Arabia | 1302 | 5.75 | 3.17 | 0.17 |
| 23 | Brazil | 1317 | 5.65 | 3.02 | 0.16 |
| 24 | South Africa | 1488 | 5.61 | 3.11 | 0.16 |
| 25 | Malaysia | 1319 | 5.59 | 3.07 | 0.17 |
| 26 | Taiwan | 1731 | 5.41 | 3.08 | 0.15 |
| 27 | Turkey | 1755 | 5.23 | 3.14 | 0.15 |
| 28 | Russia | 2203 | 5.16 | 3.34 | 0.14 |
| 29 | China | 3448 | 4.94 | 3.29 | 0.11 |
| 30 | India | 2145 | 4.62 | 3.28 | 0.14 |

## Migration

Every person has the right to leave his or her country, but the country that the person wants to move to also has the right not to accept that person. There can be several reasons why people may want to leave their own country, both positive and negative. This is often referred to as the push-pull factors in population movement.[35] Push factors can include anything from lack of food and natural disasters to political oppression.

**Refugees**

Within the push factor category are refugees or people who flee their country because of a fear of persecution. This includes individuals whose race or religion makes them targets for persecution to ethnic cleansing and armed conflict.[36] Some of the more famous examples in recent history include the first Persian Gulf War in 1991 which created a refugee population of 5 million; genocide in Rwanda in 1994 drove more than 1.7 million refugees from their homeland; and 3.5 million Afghans sought asylum from the war against terrorists in 2001.[37] Within the last five years the largest forced displacement emergency has been in Darfur in Sudan, where violence has driven hundreds of thousands of Sudanese to flee to neighboring Chad.

Often a large proportion of refugees come from countries whose governments have collapsed or they have been committing human rights abuses against their people. Kathleen Newland wrote in 1994, “Refugees know that they cannot expect, at home, the protection of the police, access to a fair trial, redress of grievances through the courts, prosecution of those who violate their rights, or public assistance in the face of disaster ... Forcing people to flee is a violation of their human right to remain peacefully in one’s home. The direct denial of other basic rights, including the rights of civilians not to be targeted in military actions, often provides the immediate impetus for flight.”[38]

The Office of the United Nations High Commissioner for Refugees (UNHCR) feels that states that once had generous refugee policies now see the costs of asylum as outweighing its benefits. In the past these refugees would arrive in controllable numbers and work jobs many others were not willing to work.[39] Since the end of the Cold War, however, many states see refugees as a burden rather than an asset. “Furthermore, since the 11 September 2001 attacks in the United States, state security concerns have come to dominate the migration debate, at times overshadowing the legitimate protection needs of individuals."[40]

In 2007, most refugees continue to live in camps similar to the one pictured previously, with no prospect of a solution. Additionally, many of the conflicts that have led to refugee flows over the last 20 years still continue without a resolution in sight. UNHCR writes, "Long-standing refugee populations have, in some cases, strained relations between host states and countries of origin. The long-term presence of refugee populations is argued to cause instability in neighboring countries and trigger intervention and insurgency."[41]

**Total population of concern to UNHCR: by category and region, 1 January 2005**

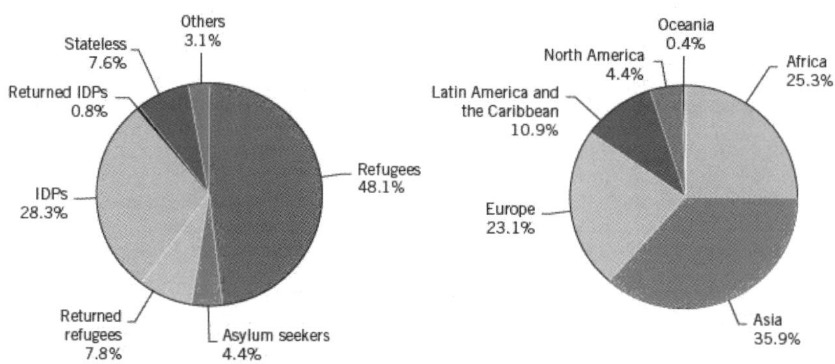


(Source: UNHCR)

*IDPs (Internally Displaced Persons) stands for people who are involuntarily uprooted from their homes but are still living in their own countries.

## Brain drain

On the opposite side of the spectrum is the pull factor. Immigrants can very often be a valuable resource to their new country. Since globalization began, one group of migrants has come under scrutiny. Often labelled 'highly skilled' migrants, these individuals from poorer countries who are highly educated in fields such as science and technology, are flocking to the industrialized world drawn by the promise of better salaries and working conditions.[42] This phenomenon is known as the 'brain drain'.

The brain drain is highly beneficial to countries such as Canada, U.K. and Australia because highly skilled foreign or foreign-born workers are a relatively important component of the professional workforce. In the U.S., it is estimated that in 1998, a quarter of Silicon Valley firms were headed by immigrants from China and India, collectively generating almost US$17 billion in sales and more than 50,000 jobs.[43] Intel, Ebay and other well-known 'American' names were in fact conceived by non-U.S. nationals.[44]

For the countries 'losing' many of their best and brightest, the emigration of highly skilled people can place a heavy burden on basic social services, including education and health. Dominique Guellec, who is responsible for science and technology indicators at the OECD, informs us that doctors' and nurses' leaving Ghana or South Africa to pursue careers abroad has proved disastrous for their home countries. Furthermore, lecturers from Africa during a March 2006 conference warned us by saying that migration of academics from Africa to universities in Europe and the U.S. is having a damaging effect on the continent's higher education. The report states, "About 30% of Africa's university-trained professionals and up to 50,000 Africans with PhDs now live and work outside the continent."[45]

Unfortunately it is very hard to know exactly how big of a problem this really is or if it is even a problem at all. Dominique Guellec explains that classifying migrants is a nightmare for statistical analysts aiming to provide an international overview. He writes, "A migrant can, for instance, be defined as a foreign national or as foreign born—two very different things. And different countries use terms referring to migrants, such as 'temporary' or 'permanent', in different ways."[46] Another major problem is that one country's methods of gathering data can differ widely from the others.

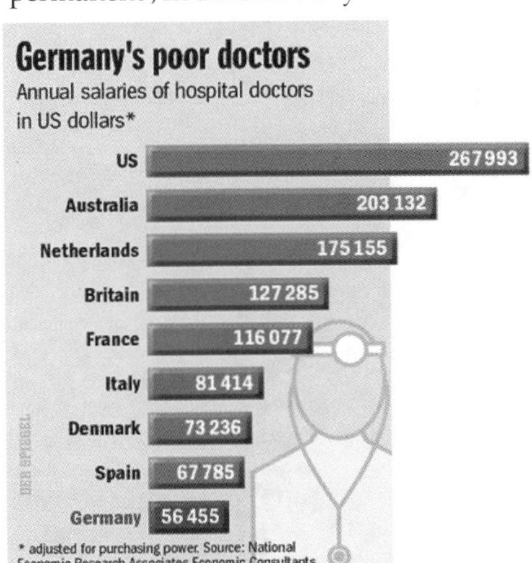


A good example comes from South Africa, where it has been very difficult to measure exactly how many people have left their home country. Crush and Williams write, "Statistics South Africa recorded a total of 62,088 people emigrating from South Africa between 1987 and 1997 to the five main destination countries (Great Britain, United States, Canada, Australia, New Zealand). However, destination country statistics of immigrant arrivals from South Africa paint a rather different picture: they show 32,296 professionals and 198,393 total immigrants arriving from South Africa during the same time period."[47]

Although this appears to be a large difference in statistics, other organizations have calculated even greater numbers. "By applying an under-counting formula for official data for the period 1989-2003, World Bank researchers have estimated total emigration for the period at 521,571 (compared to 130,965 in the official data). Overall then, between 1987 and 2003, South Africa lost an estimated 720,000 emigrants to five developed countries."[48] Nevertheless, even the World Bank numbers are understated because obviously South Africans emigrated to more than just those five countries. Furthermore, it doesn't take into account the importance of South Africans who leave for temporary contracts overseas. "While it might be argued that this is a 'temporary loss' the impact on the over-burdened and understaffed sectors within South Africa is still considerable."[49]

## Women's rights

In the past, many women around the world faced discrimination in employment and in salary. In non-Western countries, besides being denied employment they were often denied education. Women also became victims of violence and abuse in war-torn regions. During the 1990s women made some substantial gains in the areas of work and education, but problems still persist.

Iran has been one of the countries struggling to deal with the issue of women's rights. Zahra Eshraghi, the granddaughter of the Islamic revolution's leader, Ayatollah Ruhollah Khomenei, said in an interview in June of 2005 that working on women's issues has been very difficult because women did not feel safe to criticize the laws.[50] "There are certain things that are

Rally in Shaka, organized by the Jatiyo Nari Shramik trade union, an organization affiliated to the Bangladeshi trade union Kendra.

considered as crimes although the situation is gradually changing," she said. "For example it would have been very dangerous to talk about changing the constitution, or women's right to choose their dress. There can be no progress if women don't feel they are safe to express their demands."[51]

Some 89 women in Iran even chose to run for president, but were rejected on the basis of their sex. Nevertheless, a group of Iranian women activists did find the courage to force their way into the stadium to watch a soccer game between Iran and Bahrain for the first time since the Islamic Revolution banned women from watching games at the stadiums. For four hours, they carried signs that read, "My right is also human rights" and "Freedom, justice and gender equality."[52]

The sad fact remains that globally, women still face issues including violence and harassment. The most recent UN report on women's rights entitled 'Ending Violence Against Women: from Words to Action', reveals that, "a significant proportion of women suffer physical, sexual, or psychological abuse."[53] Some of the more unnerving facts are:

Murders of women often involve sexual violence, with between 40 and 70% of female murder victims killed by husbands or boyfriends in the U.S., Australia, Canada, Israel, and South Africa;

More than 130 million girls are victims of female genital mutilation, a practice most prevalent in Africa and some Middle Eastern countries but also found in immigrant communities in Europe, North America and Australia;

Female infanticide, prenatal sex selection and systematic neglect of girls were said to be widespread in South Asia, North Africa and the Middle East;

Women experience sexual harassment thoroughout their lives, with between 40 and 50% of women in the EU reporting some form of sexual harassment.

# Chapter 5

# International security

"It is only the dead who have seen the end of war."—Plato, Greek philosopher

Since the beginning of man, there has always been conflict. The main difference is we now possess weapons that can cause indescribable destruction, so much so that the average person cannot simply comprehend the damage.

Today we feel as if no one is safe. In addition, international peace and security topics seem to top the media headlines every single day. This leads people to believe that there is a strange sense of confusion in this world.

Indeed this is a confusing time. In the past, states' military strength would enable them to project power, exercise influence and dominate others. This is why states would often respond to perceived threats by arming themselves. Kegley and Wittkopf write, "Yet the capacity to destroy does not always make for a convincing threat, especially against countries and transnational terrorists with the capacity to retaliate, whose citizens, not soldiers, would die in the largest numbers. Today, the threat of force often lacks credibility especially amongst terrorist groups willing to die for their cause."[1]

Although terrorism may be the main concern for many, the fear of another huge conflict causing a world war still lingers. To combat both of these issues, states continue to spend their money on their military and other various weapons. In addition, many local wars are currently taking place that affect millions of people around the globe.

# Terrorism

Terrorism has been around for many centuries, but in the last 40 years it has changed. Since the late 1960s, more than 14,000 terrorist attacks have occurred worldwide. These international terrorist organizations have carried out increasingly destructive, high-profile attacks to call attention to their goals and to gain media coverage.[2] The first major incident occurred during the 1972 Summer Olympic Games in Munich, Germany. Since that event, few regions have been spared from terrorist attacks as terrorists will stop at nothing to destroy what they feel are evil forces.

September 11, 2001 changed the world forever. It was the event of my generation and one I find painful as I grew up close to New York City. As much as I would like to forget that day, it showed me how far terrorists are willing to go for their cause. Their radical religious and cultural motives involve violence and the targeting of innocent civilians in the places that they feel safe. The terrorists want to put fear in people and they have succeeded. Since that horrific day in 2001, the bombings have continued, most notably in Europe. Now most people are just waiting for the next larger and more devastating event to happen.

This event came very close to happening in August of 2006, when the U.S. narrowly avoided what the British called "mass murder on an unimaginable scale."[3] If the terrorist operation succeeded, the terrorists would have blown up ten U.S.-bound passenger jets with liquid explosives hidden in carry-on luggage. Thousands of men, women and children would have perished over the Atlantic.

Amanda Ripley of *Time* magazine writes, "We should be feeling safer now … but then why did it feel so bad? Why did a bullet dodged feel like the beginning of something and not the end?"[4] Bruce Hoffman, a counter-terrorism analyst at the Rand Corp, points out that, "This is the story of terrorist threats. We close up one set of vulnerabilities, and they attempt to exploit another."[5]

Many ask the question, is President Bush's post-9/11 strategy of go-it-alone and attack the enemy before they attack the U.S. working? One can argue that there have been no more attacks on U.S. soil since that day so his strategy has been successful, while others argue that Bush and his administration have just thrown more fuel into the fire, dragging his allies along with him, while eroding civil liberties at home. The overall big picture is we will never be completely safe. French terrorism expert Roland Jacquard states, "There will always be holes. One hundred percent security doesn't exist. We can do everything possible or viable to increase our security, but cutting off our arm because your hand risks gangrene is going too far."[6]

The latest U.S. State Department report on terrorism tells us that the enemy we face today in the war on terror is not the same enemy we faced on 9/11. "Today, the principal terrorist enemy confronting the United States is a transnational movement of extremist organizations, networks, and individuals—and their state and non-state supporters—which have in common that they exploit Islam and use terrorism for ideological ends."[7]

The State Department also informs us that the terrorists are fueled by a radical ideology and a false belief that the U.S. is the main cause of most problems affecting Muslims today. The scary part is that many people believe it. Even scarier is the fact that many states like the old Taliban-ruled Afghanistan, have served as sanctuaries for extremists to launch additional attacks against not only the U.S., its allies and partners, but the Muslim world itself. The terrorists seek to expel Western power and influence from the Muslim world and establish regimes that rule according to a violent and intolerant distortion of Islam.[8] One of the worst possible scenarios and the goal of many terrorists is highlighted in the report. It states, "Some among the enemy, particularly al-Qaida, harbor even greater territorial and geopolitical ambitions and aim to establish a single, pan-Islamic, totalitarian regime that stretches from Spain to Southeast Asia."[9]

Again, for the terrorist, there is no peaceful coexistence with those who do not accept their radical view. They use beheadings, suicide bombings, and other horrible acts as a means to promote their belief. Undoubtedly the gravest danger would be if these individuals were able to get hold of weapons of mass destruction (WMDs). The State Department writes, "Our enemy's demonstrated indifference to human life and desire to inflict catastrophic damage on the United States and its friends and allies around the world have fueled their desire for weapons of mass destruction."[10]

## Weapons of mass destruction

WMDs have different definitions but most refer to the ABCs, meaning atomic, biological and chemical weapons. Terrorists have been in possession of WMDs in the past and they haven't been afraid to use them. The most notorious act of biochemical terrorism was the sarin gas (a deadly nerve gas) attack in the Toyko subway in 1995. It was committed by a radical religious cult that wanted to control Japan.[11] Twelve people were killed and more than 5,700 were injured.

Chemical and biological weapons remain the biggest threat. They can inflict extreme casualties but also create just as much fear. Furthermore, they are often invisible and odorless to human senses so before you even know an attack has been carried out, it is too

America's Titan II missile, which was made ready to launch at a moment's notice during the Cold War.

late. Another reason why they are such a big threat is because biochemical agents are very easy to acquire. Most of the agents needed to make a biological or chemical weapon can be found in your local stores.

A nuclear terrorist attack is a lot less likely than that of biological or chemical, but terrorist organizations are trying to get a hold of the materials necessary to make such a weapon. The most likely source of such materials would come from rogue states such as Iran.[12] However, a more plausible scenario would be one state attacking another with a nuclear weapon. In the post-Cold War era, this type of event would also appear to be highly unlikely, but today various countries have the technology to produce such weapons.

In October 2006, North Korea successfully detonated its first atomic weapon. This has worried the entire world, especially the U.S. which the North Koreans see as their main enemy. The following week, Mohamed ElBaradei, head of the UN's international Atomic Energy Agency (IAEA), warned that as many as 30 countries could have the technology to produce atomic weapons "in a very short time." The article states, "ElBaradei did not single out any country in his warning, but was clearly alluding to Iran and other nations that are working to develop uranium-enrichment capability, such as Brazil."[13]

There are five formally declared nuclear weapon states: the U.S.; Russia; China; France; and the U.K..[14] Four others are believed or known to have such arms: India, Pakistan, Israel and now North Korea. Canada, Germany, Sweden, Belgium, Switzerland, Taiwan, Spain, Hungary, the Czech Republic, Slovakia, and Lithuania could realistically build an atomic device quite quickly, if they chose to, but none has shown an interest in doing so. Japan and South Korea are two others who could use their plutonium left over from their nuclear power plants to construct a nuclear weapon.[15] According to the UN, others considering developing nuclear programs are Egypt, Bangladesh, Ghana, Indonesia, Jordan, Namibia, Moldova, Nigeria, Poland, Thailand, Turkey, Vietnam and Yemen.

In October of 2006, the U.S., which has the largest nuclear arsenal in the world, announced the start of a multiyear process to repair and replace facilities where nuclear weapons have been developed and assembled and where older warheads could be more rapidly dismantled.[16] This is just one step of many toward building a new stockpile of up to 2,200 deployed nuclear weapons that will last for most of the 21st century. This is a significant drop from the Cold War stockpile of about 6,000 warheads, but it is still a far cry from eliminating their entire nuclear arsenal, as many would like to see done.

## Small arms and light weapons

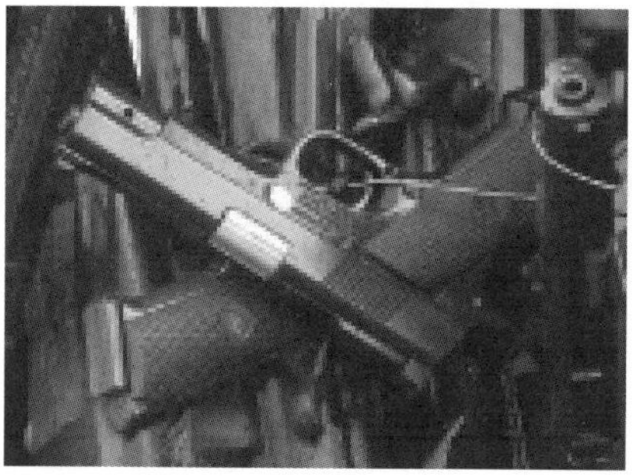

Small arms and light weapons (SALW) are what many consider to be the true WMDs. There are about 639 million SALWs in the world today.[17] Eight million more are produced every year. Each year, at least a third of a million people are killed directly with these weapons.

Virginia Gamba of the international organization SaferAfrica informs us that certain groups in society and government benefit from lawless conditions and involvement in illicit SALW trafficking. When the citizens obtain these weapons the usual result is armed violence and crime. This can have implications such as slow economic recovery, discourage investment, and threaten the development of any economic enterprise from national to local levels. Gamba writes, "A critical spin-off from these problems is erosion of community confidence and mistrust between individuals and the authorities—which has long-term implications for the legitimacy of the government."[18]

If SALWs are not controlled, they will continue to fuel violent conflict, state repression, crime, and domestic abuse. Controlarms.org works hard to send out the message that unless governments act to stop the spread of arms, more lives will be lost, more human rights violations will take place, and more people will be denied the chance to escape poverty.[19] Many organizations have worked extremely hard in the past, but now is the time to take some serious action. "Urgent measures are needed immediately."[20]

## Today's wars

SALWs play an enormous role in the wars taking place at this very moment. Since globalization, the global arms trade has changed dramatically. Arms companies are now located all over the world. Furthermore, their products are often assembled in countries that do not worry where the weapons end up.

SALWs are not the only problem as the overall arms trade has changed since globalization. The October 2006 Control Arms Campaign report entitled 'Arms without Borders' provides many case studies that illustrate aspects of the changing industry and the inadequacy of current law to control it. The main point here is that often SALWs and other arms end up in the hands of governments with track records of using weapons inappropriately and unlawfully against civilians in violation of international human rights law and humanitarian law.[21] When this occurs governments can easily control their people, which in turn can spark civil war. The following is a list of the ongoing wars taking place today and a list of where they are located in the world:

| Start of conflict | War/conflict | Location |
|---|---|---|
| 1964 | Colombian armed conflict | Colombia |
| 1969 | New People's Army/ Islamic insurgency (including OEF-P) | Philippines |
| 1983 | Sri Lankan civil war | Tamil Eelam |
| 1984 | Kurdish separatist insurgency | Turkey and Kurdistan |
| 1984 | Free Papua Movement | Western New Guinea |
| 1988 | Casamanace conflict | Senegal |
| 1988 | Somali civil war | Somalia |
| 1989 | Kashmir conflict | Kashmir |
| 1993 | Ethnic conflict in Nagaland | Nagaland, India |
| 1996 | Nepal civil war | Nepal |
| 1999 | Ituri conflict | DR Congo |
| 1999 | Second Chechen war | Russia |
| 2000 | Al-Aqsa intifada | Israel/Palestinian territories |
| 2000 | Laos conflict involving the Hmong | Laos |
| 2001 | Post-Taliban war in Afghanistan | Afghanistan |
| 2001 | South Thailand insurgency | Pattani (region) |
| 2002 | Ivorian civil war | Côte d'Ivoire |
| 2003 | Iraq war | Iraq |
| 2003 | Balochistan conflict | Balochistan (region) |
| 2003 | Darfur conflict<br>Chadian-Sudanese conflict<br>Second Chadian civil war | Sudan/Chad |
| 2005 | Western Sahara independence intifada | Western Sahara |

(Source: http://en.wikipedia.org/wiki/Ongoing_wars)

## Military spending

Military spending around the world is ridiculously high. Control Arms Organization points out that if the current growth in worldwide military spending continues, it will pass the highest figure reached during the Cold War. In 2006, global military spending went over the one trillion US$ mark, which is roughly 15 times the annual international aid expenditure.[22] In 2005, it was estimated that global spending on arms alone was 34% higher than in 1996.[23] These figures show us that the post-Cold War decline in military spending is long gone.

Global military spending is increasing and expanding for several reasons, one being the global arms trade. In addition, many key reasons are the war on terror and the increased spending by rising economies, both showing no signs of slowing down.[24] There is also a trend for states that cannot afford it, to spend huge amounts on their military. Controlarms.org writes, "Some of the poorest countries in the world, including Botswana, the Democratic Republic of Congo, Nigeria, Rwanda, Sudan and Uganda, are among those that doubled their military spending between 1985 and 2000. In 2002-03, Bangladesh, Nepal and Pakistan were among those governments that spent more on their military than on healthcare. In some developing countries, high military spending bears little relation to real defence needs. In Angola, for example, the proportion of GDP devoted to military spending more than doubled in the two years *after* its 27-year-long war ended in 2002, rising to 4.2%."[25]

In other countries, government arms purchases alone can exceed legitimate security needs, diverting large amounts of money from programs that really need it. The U.S. Congressional Research Service estimated that collectively, countries in Asia, the Middle East, Latin America and Africa spent $22.5bn on arms during 2004; 8% more than they did in 2003.[26] This sum of money would have enabled those countries to put every child in school and to reduce child mortality by two-thirds by 2015.[27]

Looking at the bigger picture, the U.S. is by far the number one spender. For the fiscal year (FY) 2007, the U.S. military budget is $462.7 billion.[28] Anup Shah reports that this figure includes the Defense Department budget, funding for the Department of Energy (which includes nuclear weapons) and a category labeled 'other' which this source does not define. However, "it does not include other items such as money for the Afghan and Iraq wars—$50 billion for FY 2007 and an *extra* $70 billion for FY 2006, on top of the $50 billion approved by Congress."[29] Examining the U.S.'s national debt proves that it is not only poor countries who cannot afford to continue this trend of expensive military buildup.

**Global priority table:**

| Item | $US billions |
|---|---|
| Cosmetics in the U.S. | 8 |
| Ice-cream in Europe | 11 |
| Perfumes in Europe and U.S. | 12 |
| Pet foods in Europe and U.S. | 17 |
| Business entertainment in Japan | 35 |
| Cigarettes in Europe | 50 |
| Alcoholic drinks in Europe | 105 |
| Narcotic drugs in the world | 400 |
| Basic education for all | 6 |
| Water and sanitation for all | 9 |
| Reproductive health for women | 12 |
| Basic health and nutrition | 13 |

(Source: www.globalissues.org/Geopolitics/Armstrade/Spending.asp)

*Note*: Some countries are excluded due to lack of data or consistent time series data. World totals exclude Afghanistan, Angola, Benin, Cuba, Equatorial Guinea, Haiti, Iraq, Liberia, Myanmar, Qatar, Somalia, Trinidad, Tobago and Vietnam.

Author Chris Hellman notes: "Linking military spending to the GDP is an argument frequently made by supporters of higher military budgets. Comparing military spending (or any other spending for that matter) to the GDP tells you how large a burden such spending puts on the U.S. economy, but it tells you nothing about the burden a $440 billion military budget puts on U.S. taxpayers. Our economy may be able to bear higher military spending, but the question today is whether current military spending levels are necessary and whether these funds are going towards the proper priorities. Further, such comparisons are only made when the economy is healthy. It is unlikely that those arguing that military spending should be a certain portion of GDP would continue to make this case if the economy suddenly weakened, thus requiring dramatic cuts in the military."[30]

## Military technology

Just as technology continues to advance in everyday life, military technology usually progresses even faster. As we read in Chapter 1, most of our modern advances were originally developed for either the military and/or NASA. We must not forget that some of the most amazing machinery that the U.S. military has today is actually decades old. Take for example the F-117 Stealth Bomber pictured overleaf. It was developed and first flown in 1978,

almost 30 years ago. This makes it hard to even imagine what the U.S. and other militaries might possess today.

Many people who work in the industry have written about military weapons that are very close to being completed or might be operational already. J. Douglas Beason, author of the recently published book *The E-Bomb: How America's New Directed Energy Weapons Will Change the Way Wars Will Be Fought in the Future*, was the president's science adviser under both the Bush and Clinton administrations.

His book informs us that, "After more than two decades of research, the United States is on the verge of deploying a new generation of weapons that discharge beams of energy, such as the Airborne Laser and the Active Denial System, as well as the Tactical High Energy Laser, or THEL."[31] The U.S. Army has tested the laser and found that it can blast high-flying rocket warheads with enough heat and energy, which causes them to self-detonate. THEL uses a high-energy, deuterium fluoride chemical laser that has also proven to take out multiple mortar rounds as well.[32] There is also something called Active Denial Technology. It can stop an advancing army by using millimeter-wave electromagnetic energy. "This technology, supported by the U.S. Marines, uses a beam of millimeter waves to heat a foe's skin, causing severe pain without damage, and making the adversary flee the scene."[33]

Perhaps the biggest change in technology has been the use of remote-control robotics. The war on terror has introduced us to the unmanned aircraft. Most of us have seen these remote-control planes on the news as they keep a careful eye on the Special Forces and other operations in Afghanistan.

Today in Iraq you will find what are commonly called 'robot troops'. In 2005, 18 one-meter-high robots equipped with cameras and operated by

AFR: A laser fires from space toward Earth in this artistic rendering. The Air Force Research Laboratory's Directed Energy Directorate is conducting research in a wide variety of laser weapons technologies.

remote control, first arrived in Iraq. The 'robot warrior' can be fitted with an M249 or M240 rifle, can travel over rocks and barbed wire, and has four cameras, night-vision and zoom lenses.[34] A human, through the cameras and two joysticks, operates the machine and the total cost of the unit is US$200,000.[35] The robot was built based on one already being used by the military to disable bombs. Testing has shown that the robot is fast and best of all, reduces risk to the lives of soldiers.

Testing has also shown that the robot is a more accurate shot compared to the average soldier because it is mounted on a stable platform and takes aim electronically, according to a U.S. military officer. "It eliminates the majority of shooting errors you have," said Staff Sergeant Santiago Tordillos.[36] Furthermore, there are plans to replace the computer screen, joysticks and keypad in the remote-control unit with a Gameboy-style controller and virtual-reality goggles.[37]

Another new military technology is a robotic security system that can detect intruders, assess their intentions, and even stop them if necessary. The Mobile Detection Assessment and Response System (MDARS) was originally conceived to provide security at installations within U.S. borders, but officials now intend to send it to forward operating bases around the world.[38]

MDARS' purpose is to perform random patrols within assigned areas on bases and warehouses. The robot's systems are programmed to detect everything from intruders to fires. Specifically, the mobile platforms used by the machine utilize an array of sensors that are freely controlled by a monitor panel. The mobility base is diesel powered with four-wheel hydrostatic-drive and equipped with active laser, ultra-sonic sonar, millimeter-wave radar and stereo video sensors for collision avoidance.[39] The vehicle also has intrusion detection and assessment forward-looking infrared, radar, c amera and pan/tilt turret, as well as vehicle communications link antennae and differential GPS.[40]

Lieutenant-Colonel Eugene Stockel, product manager for U.S. Army force protection systems, states, "If it is too big, it will stop. It will navigate around the obstacle, or send signals to the control panel."[41] The robot can detect

Launching of the Dragon Eye Unmanned Aerial Vehicle. The Dragon Eye is a small plane guided by computers and provides real time video of the terrain below it. The U.S. Marines use the vehicle during their patrols to find any Improvised Explosive Devices or suspicious people traveling the roads.

objects right in front of them and obstacles to about 30 meters away. "It is mounted low, in front of the vehicle, to pick up the elements that fall out of view, scanning in front of the vehicle ... The robot moves to an over-watch position and commences scanning and detection," said Stockel. Moreover, it provides both day and night assessment of what is taking place. "When we get it right it is a very capable and beneficial capability for performing the drudgery of surveillance," he said.

This new technology is very frightening. Even scarier is the length some states and terrorists are willing to go so they can acquire such devices. So scary in fact that on January 17, 2007, the Board of Directors of the Bulletin of the Atomic Scientists at the University of Chicago changed their world-famous Doomsday Clock to five minutes to midnight.[42] On the clock midnight represents destruction by nuclear war. For the first time ever, experts added climate change to the prospect of nuclear annihilation as the greatest threats to humankind.[43]

Will the situation change in this century? Can the world discover other paths to peace besides arming themselves? In 1995, Martti Ahtisaari, President of Finland said, "To deal with the great security challenges of our time, including population growth, the spread of weapons of mass destruction, crime, environmental degradation and ethnic conflicts, we must resolutely adopt new methods of managing change and building global security."[44] Twelve years later, are we moving forward?

# Part 2

## Are we moving forward?

# Chapter 6

# The individual contribution

"During my lifetime I have dedicated myself to this struggle of the African people. I have fought against white domination, and I have fought against black domination. I have cherished the ideal of a democratic and free society in which all persons live together in harmony with equal opportunities. It is an ideal which I hope to live for and to see realized. But if needs be, it is an ideal for which I am prepared to die."
—Nelson Mandela

Statue of Nelson Mandela in Nelson Mandela Square, Sandton, South Africa.

After reading about the world's problems in the previous chapters, it is important we examine the various people who are trying to change the world for the better.

## Introduction—what difference can we make as individuals?

Although Nelson Mandela spent 27 years of life behind bars, what he accomplished for both Africa and the world is extraordinary. He never asked himself the question above; he simply dedicated himself to the struggle. It was Mandela who facilitated a peaceful transition to a fully representative democracy in South Africa and became the country's first black state president. Since the end of his presidency Mandela has continued his contributions to the world arena. He has lent his support to the 46664 AIDS fundraising campaign, named after his prison number. Mandela has also expressed his support for the international Make Poverty History movement of which the One Campaign is a part.[1]

The Make Poverty History campaign (which is written as **MAKE**POVERTY**HISTORY**) was a British and Irish coalition of charities, religious groups, trade unions, campaigning groups and celebrities who mobilized around the U.K.'s prominence in world politics in 2005 to increase awareness and pressure governments into taking actions towards relieving absolute poverty. The symbol of the campaign was a white wristband made of cotton or silicone. A 'virtual' white band was also available to be displayed on websites.

The ONE Campaign took place in the U.S., being their version of Make Poverty History. In April of 2005, a commercial began airing with several celebrities in black and white stating the pledge of the American ONE Campaign. The commercial featured 33 celebrities and personalities. In the end, actor Tom Hanks states, "We're not asking for your money. We're asking for your voice."[2]

After looking at only the life of Nelson Mandela, you can no longer say that one person cannot make a difference in this world. It may not be on the same scale as Mandela's, but it is a difference nonetheless. For example, do you remember the time before recycling became popular? Everyone simply threw everything in the trash. In the past 25 years that has changed. More people now recycle plastics, glass and newspapers. Today, many businesses recycle paper and buy recycled products, and many industries practice source reduction in their packaging efforts. An entire mindset has changed in one generation and a little individual action like recycling can help change the world.[3]

Going further back in history we can find even more examples of individual actions making a difference. During the time of the Holocaust everyone had to make moral choices. Some people chose to become perpetrators and others chose to be bystanders. However, a small minority of individuals chose to help the persecuted—these were the rescuers and helpers.[4] This was an extraordinary selfless choice and many paid with their lives. Although none succeeded in halting the Holocaust, many Jews were enabled to survive as a result of the efforts of these individuals. Each chose to defy the power of the Nazis and their collaborators—mostly single-handedly. Their choice made a huge difference to many individual lives and the world around them, but more importantly they showed the power that one individual could have.

U.S. Representative John Murtha is a strong believer in the difference individuals can make. As a commencement speaker for a college graduation Murtha stated, "There are lots of cynics in the world, but don't ever let them convince you that you can't make a difference. You can make a difference. The things that you do can have a tremendous impact on the lives of others. If you're a computer programmer or a nurse, a diplomat or an outspoken citizen, you can affect our productivity or our health, our freedom or our communities."[5] He went on to state, "I would urge you all to be engaged in our communities and to continually seek innovative solutions to the problems that we face ... because if you do, you will make a difference in this world."[6]

Today, many people from various countries are making a positive impact on the planet and its people. Let's take a closer look at some of the more familiar faces and examine what they are doing to make the world a better place.

## Celebrities working with the UN

Many celebrities have joined forces with the UN to help spread the word of international peace. This idea of UN celebrity ambassadors began in the 1950s when Hollywood and Broadway star Danny Kaye volunteered to promote children's rights as the spokesperson for UNICEF.[7] He worked tirelessly throughout the year and the subsequent decades. It was his success that led to the use of stars, like Sir Peter Ustinov and Audrey Hepburn, on a regular basis.[8] Today the list includes stars such as former basketball player Magic Johnson and actress Nicole Kidman.

Most of the famous faces choose to focus on one particular area like Muhammad Ali, who is a relentless campaigner for the poor. The individual UN agencies (e.g. UNESCO) also have their own goodwill ambassadors. Thus the UN Development Program or UNDP, uses Brazilian footballer Ronaldo to highlight the plight of the poor. Former Spice Girl Geri Halliwell has also famously done her bit for the UN Population Fund, promoting safe sex to teenage girls in the Philippines.[9]

Famous Brazilian soccer player Ronaldo campaigns against poverty.

Each agency has different 'ranks' of celebrities who contribute varying amounts of time and effort. For example, UNICEF, the UN Children's Fund, now has over a dozen Goodwill Ambassadors, including Roger Federer and David Beckham.[10] Others become spokespeople for specific projects. For example, Sir Roger Moore, better known as the former James Bond, is the major figurehead for the UNICEF initiative to eliminate iodine deficiency disorders (IDD).[11]

Although there are dozens of celebrities who work hand in hand with the UN to contribute to global causes, there are a certain few who I believe deserve to be singled out.

## Angelina Jolie

"We cannot close ourselves off to information and ignore the fact that millions of people are out there suffering. I honestly want to help. I don't believe I feel differently from other people. I think we all want justice and equality, a chance for a life with meaning. All of us would like to believe that if we were in a bad situation someone would help us."—Angelina Jolie

I am sure no one has forgotten the madness that took place when Angelina Jolie and Brad Pitt were first linked together romantically. Almost everyday their pictures and stories were broadcast all over the world. Although I felt awful for the 'invasion of privacy' they had to endure, one positive did come out of all this: it highlighted Ms Jolie's role as a UN High Commissioner for Refugees (UNHCR) Goodwill Ambassador.

Jolie first became aware of worldwide humanitarian crises while filming the movie *Tomb Raider* in Cambodia. In an early interview she stated, "I discovered things about what's happening in the world ... Cambodia was really eye-opening for me."[12] Due to these experiences she went to UNHCR for more information. In the following months after the end of filming, she visited different refugee camps around the world to learn more about the situation and the conditions in the area. From February to August 2001 Jolie went on missions to Sierra Leone, Tanzania, Cambodia, and Pakistan. In addition, she insisted on covering all costs related to her missions and shared the same rudimentary working and living conditions as UNHCR field staff.[13]

Impressed by her interest and devotion, UNHCR named her a Goodwill Ambassador on August 27, 2001 at its headquarters in Geneva.[14] Since then she has worked in many different areas around the world. Her work includes:

- Visiting a Osire refugee camp in Namibia while filming the movie *Beyond Borders*. She and the production team later donated 270 tents and several hundred items of bedding and mattresses to the camp.

- Visiting the Tham Hin refugee camp in Thailand.
- A four-day trip to Ecuador visiting Colombian refugees.
- Attending a landmine awareness center in Cheshire, U.K..
- Visiting the Kakuma refugee camp in Kenya that hosts more than 80,000 refugees, mainly from Sudan.
- A four-day mission to Kosovo, visiting return sites and meeting with a team of mine clearing women in the capital.
- A six-day mission to Tanzania and traveling to the western border camps to learn more about UNHCR's operations in the area. She visited 85,000 Congolese refugees, and followed a group of 91 separated children who had just arrived from the DRC, while donating US$50,000 to the Kurasini orphanage in the capital.
- Paying a week-long visit to Sri Lanka to get a first–hand look at the post-war conditions in northern Sri Lanka, visiting recent returnees, IDPs, and other victims of the 20–year civil war.
- A four-day mission to Russia which included learning of UNHCR's operations in the region and meeting displaced Chechens in their tent homes.
- A private stay in Jordan where she visited a camp in the country's remote eastern desert, 70km from the Iraqi border.
- Visiting Sudanese refugees near the capital of Egypt. She helped distribute winter clothes, blankets and toys to Sudanese refugees living there and also donated US$20,000 to a community health project in the area.
- Visiting detained asylum seekers at three facilities in Arizona as part of her efforts to give a voice to the tens of thousands of asylum seekers detained in the United States.
- Visiting Chad, paying a visit to border sites and camps for refugees who had fled fighting in Darfur. At the end of the trip, she emphasized the urgent need for UNHCR and its partners to get additional funding in order to continue assisting the Sudanese refugees.
- Visiting UNHCR's regional office in Beirut, Lebanon, as well as some young refugees and cancer patients in the capital.
- Visiting different Pakistani camps harboring Afghan refugees while also meeting with Pakistan's President Pervez Musharraf and Prime Minister Shaukat Aziz.
- Returning to Pakistan with Brad Pitt to see first-hand the impact of the 2005 earthquake.
- Visiting Haiti with Brad Pitt and supporting Yele Haiti, a charity founded by Haitian-born Wyclef Jean.[15]

Her extensive work has landed her many awards. Cambodia's King Norodom Sihamoni awarded Angelina Cambodian citizenship for her work in the country. She has also pledged US$5 million to set up a wildlife sanctuary in the northwestern province of Battambang and owns property there.[16]

Jolie has also done many other things besides those in the list above. In 2003, she published a collection of journals made during missions for the UNHCR. Entitled 'Notes from My Travels,' the collection of journal entries chronicle her early field missions for UNHCR. More importantly, all the proceeds from the book went to the UNHCR.

Additionally, Jolie filmed a MTV special entitled, 'The Diary of Angelina Jolie & Dr. Jeffrey Sachs in Africa'. portraying her and noted economist Dr. Jeffery Sachs on their trip to western Kenya. There, Sachs's UN Millennium Project team is currently working with locals to end poverty, hunger and disease.[17]

**Continued support—world education and children**

Throughout 2005 and 2006, Jolie continued her fight for the betterment of world education and children everywhere. In March of 2005, she took part in a Washington Press Club luncheon where she announced the founding of the National Center for Refugee and Immigrant Children. It is based in Washington, DC, and is an organization that provides free legal aid to children who prior to this had no legal representation. Angelina donated US$500,000 to the center, which will help keep it afloat for the first two years of its operation.[18]

In November 2005, Jolie called on the U.S. Congress to fund global children's programs and eliminate school fees. "America has the opportunity to lead the world in addressing the AIDS orphan crisis," she said in a Capitol Hill press conference. "We all know we should do this. I hope that we will." She also highlighted the importance of guaranteeing universal access to quality education. She stated, "New medicines are giving children with AIDS a chance at life, so what are they supposed to do with that life? Sit idle? Most of them have lost their parents to AIDS and they're trying to raise themselves, they're trying to feed themselves, and now they have to send themselves to school. We must eliminate school fees."[19]

In 2006, in honor of Global Education Week, NBC's Ann Curry was granted an exclusive interview with Jolie and her new cause, calling for all the world's children to be given the opportunity to get an education. She stated that "Lack of education causes death ... More children die under the age of five when the parents are not educated. More people get AIDS when they haven't had an education. Statistics prove that if every child was in school every year, 700,000 less people would get AIDS."[20]

Angelina Jolie with U.S. Secretary of State Condoleezza Rice.

Curry also asked why the American

people should care about the issue when there are so many issues to worry about at home. She answered, "Because if we just fix our home and the neighborhoods around us fall apart ... what kind of a place are we going to live in?"[21]

There is no doubt that Jolie will continue her quest to making the world a better place. In the past, UNHCR's representative to Kenya, George Okoth-Obbo, praised her "presence, just to bring some joy into what is undoubtedly a hard life for many of the people here."[22] I am sure she will continue to bring joy to many peoples' lives for years to come.

### Michael Douglas

"Wherever arms flow, violence follows."—Michael Douglas

Born in Middlesex County, NJ, USA, actor Michael Douglas is one of the eight UN Messengers of Peace. The Messengers of Peace are distinguished men and women of talent and passion, who have expressed their desire to help focus global attention on the noble aims of the UN Charter: a world without war, respect for human rights, international law and social and economic progress.[23]

Douglas, whose interest in social justice and peace issues dates back to his participation in *The China Syndrome* of 1979, was appointed a Messenger of Peace by then Secretary-General Kofi Annan in 1998. As a Messenger of Peace, he has focused his advocacy on disarmament, including nuclear non-proliferation and stemming the spread of small arms, as well as on human rights.[24] In 2006, he was featured in a public service announcement (PSA) to focus attention on the issue of the illicit trade in SALW. The PSA was to promote the UN Conference to Review Progress Made in the Implementation of the Program of Action to Prevent, Combat and Eradicate the Illicit Trade in Small Arms and Light Weapons in all its Aspects which took place at UN Headquarters from June 26 to July 7, 2006.

The PSA was an integral part of a UN public information campaign, and called attention to the dangers posed by the illicit trade in SALW, including their humanitarian and social impact on children and the civilian population in general. Mr. Douglas states that "The conference is an opportunity for UN member states to build on the Programme of Action and to encourage countries to strengthen their laws on the illicit trade."[25]

The spot emphasizes that the illicit trade in SALW is "an issue that affects us all." It also reinforces the point that, "while owning guns is a legal right in most countries, the illegal trade in guns continues to fuel conflict, crime and violence." Mr. Douglas' appearance in the PSA reflects his strong commitment to disarmament and his willingness to further this message through the power of television and the Internet.[26]

Three years earlier in 2003, Mr. Douglas hosted a powerful film on child soldiers which explored the impact of combat on children in Sierra Leone. During the film you meet a boy named Abu who explains to Douglas that he doesn't know how old he is. Abu is one of an estimated 300,000 children worldwide who have been conscripted or kidnapped to fight as child soldiers. "After being kidnapped by a rebel group, he was tortured, drugged, and forced to commit atrocities."[27]

Throughout the 30-minute film, Abu searches for his long-lost family. After several fruitless searches in one part of the Sierra Leone, he decides to follow one last lead into another region of the country. With Michael Douglas in hand, they fly out on a UN helicopter and walk for miles along a dusty road from one village to the next. Finally, while waiting to make inquiries of the chief in a remote village, Abu hears a cry of joy. It is his mother.[28]

"It's incredible to see Abu in his mother's arms. I'm overwhelmed," says a visibly moved Douglas in the film. "I never expected to see Abu reunited with his family."

Speaking about his role as a Messenger of Peace, Douglas states, "I'm in an enviable position ... When I talk about movies I can talk about messages of peace, and infuse them into the entertainment pages."[29]

## Other celebrities

Although the following people are not affiliated with the UN, it doesn't mean what they do is more or less important than the others. The bottom line is that all the people in this chapter have made a substantial impact on global issues. Moreover, one of the main points I want people to take away from this chapter is the fact that it doesn't take much to make a difference. So, let's take a look at even more good deeds performed by people from very different backgrounds.

### Oprah Winfrey

"You get from the world what you give to the world."—Oprah Winfrey

Everyone knows Oprah. It is that simple. Her television show, *The Oprah Winfrey Show* has been on for years and is loved throughout the world. What has evolved out of the television show is the marvelous organization called Oprah's Angel Network.

On the September 18, 1997 episode, Oprah invited viewers to join her in using their lives to improve the lives of others. The campaign was started simply

to encourage viewers to collect spare change for 150 scholarships given through The Boys & Girls Clubs of America and to volunteer time to build 200 homes for the needy with Habitat for Humanity. However, it eventually evolved into the charity known today as Oprah's Angel Network.[30]

In 1998, Oprah's Angel Network was formally established as a public charity. Its goal is to encourage people around the world to make a difference in the lives of others. Oprah's vision is to inspire individuals to create opportunities that help underserved people rise to their own potential.[31] Her Network supports charitable projects and provides grants to non-profit organizations across the globe that share in this vision.

Oprah's Angel Network has been involved in building rural schools in ten different countries (China, Ecuador, Ghana, Guatemala, Haiti, India, Mexico, Nicaragua, Sierra Leone and Tanzania). Thirty-four schools have been built through Kids Can Free the Children, providing education for thousands of children in remote areas of the world. The organization has also raised more than US$50 million, including US$9 million from viewers responding to the televised coverage of the Christmas Kindness South Africa 2002 initiative and more than US$10 million for Hurricane Katrina victims.[32] On top of this, Oprah Winfrey is generous enough to make donation to the Network equal to its management.

Some of the organization's achievements so far include:

- Distributing 16,000 new uniforms, shoes, and school supplies to South Africa's most vulnerable children;
- Providing support for community-based organizations in South Africa, Zimbabwe and Zambia that are helping parentless children thrive;
- Building a new school to serve some of the Christmas Kindness children whose school was recently torn down, as well as close to a dozen rural schools in other developing nations; and
- Supporting groups that help women in post-conflict areas become productive and regain their dignity.[33]

**Oprah and South Africa**

Being a permanent resident of South Africa, I was very excited to learn about the work Oprah was doing for the country. Starting in 2002, Oprah and the Network held three Christmas Kindness parties for orphaned children who lived in townships from Soweto, Durban and Umtata. Employees from the organizations of HOPE worldwide and the God's Golden Acre orphanage helped them gather some of the neediest children affected by AIDS. These children had no food, no family, and no home, but because of Oprah and her viewers, they had a Christmas celebration that they will never forget.[34] In an article from June 13, 2005 entitled 'Spirit and hope in SA: Oprah', Mary Alexander informed us that Oprah believes that South Africans have more spirit and hope than any other people she has ever met. In June 2005, Oprah was in South Africa to give a speech at the Live Your Best Life seminar at the Sandton Convention Centre in Johannesburg.[35] Also while in country, Winfrey donated R160,000 to the Centre for the Study of Violence and Reconciliation, an NGO working towards peaceful and fundamental transformation in South Africa, and about R6 million to the Ithuteng Trust, a school for reformed delinquents in Soweto. On top of that, Oprah opened a R175-million premier girls' school, the Oprah Winfrey Leadership Academy for Girls, on nine hectares of land at Henley-on-Klip in Gauteng. She stated that "I want to create the premier school for girls in the world—one I'd like to have gone to."

**HIV/AIDS and poverty**

Oprah stated that Nelson Mandela had told her he believed Princess Diana had done a lot of good by being seen hugging AIDS babies, and that celebrity attention in the U.S. had changed the way people viewed the pandemic. Oprah also said she was glad she'd been poor and was looked after by her grandmother because that was one reason she empathized so much with South African children, who were often in the same boat.[37]

"When I was ten, I saw Sidney Poitier win an Oscar," she said. "It made me wonder what I could do. The same thing happened when I met a 16-year-old girl in Soweto this week who is the head of a household of six." Oprah believes poverty can be eradicated in our lifetime. "It exists only because the world is not outraged about it," she said. During her tour of South Africa, Oprah joined in traditional Zulu dancing on a visit to Soweto, reports *The Star*, a South African newspaper. Winfrey couldn't contain herself as hundreds of pupils from Ithuteng Trust—a school for reformed delinquents in Klipspruit—welcomed her.[38]

Oprah dedicated the day to the less fortunate children of Soweto, first visiting AIDS orphans in Soweto and then in Zuurbekom outside the township. She sat and talked with the children's caregivers and helped distribute school uniforms. At Ithuteng, 900 schoolchildren—some reformed gangsters and drug addicts, others victims of abuse—sang songs and toyi-toyied, to Oprah's delight. A play performed by the children depicted domestic violence and rape. She couldn't control her tears. Her bodyguard was seen passing her tissues and sunglasses. She even tried to perform the *ukusina*, a traditional Zulu dance.[39]

The children also gave her traditional ornaments, including a Basotho grass-woven hat, a gourd and a grass mat. Oprah shared her history of poverty and abuse, revealing that she was raped by three family members. The last rape, when she was 13, resulted in a child, who later died. "I stand with you and understand where you come from," she said. "All things are possible; there is nothing you cannot do. What I want you to do here is work on yourself."[40]

## Bill Gates

"I do think this next century, hopefully, will be about a more global view. Where you don't just think, yes my country is doing well, but you think about the world at large."
—Bill Gates

Most of us know Bill Gates as the founder of Microsoft and the richest man in the world, but most of us are not aware of the fact that his foundation may arguably be the most important charitable organization in the world. Officially titled 'The Bill & Melinda Gates Foundation,' ('B&MGF,' named after Bill and his wife Melinda), is currently the largest transparently operated charitable foundation on the planet, and as of June 2006, it has an endowment of approximately US$29.2 billion.[41] The main aim of the foundation is to provide lifesaving healthcare products and technology for the poorest parts of the world.

The foundation is of such dear importance to Bill Gates and his wife that on June 15, 2006, he announced his plans to transition out of a day-to-day role with Microsoft. This move will be effective as of July 31, 2008, as it allows him to devote more time to working with the foundation.[42]

Other important news came during the month of June in 2006 when Warren Buffett, the world's second-richest person after Gates, pledged to give the foundation approximately ten million Berkshire Hathaway Class B shares (worth US$30.7 billion as of June 2006) spread over multiple years through annual contributions.[43]

I know what you are thinking, 30 billion dollars is an insane amount of money just to hand over to a foundation. This is indeed true, and because of it, Buffett set conditions on his money. Buffett's contributions do not simply increase the foundation's endowment, but effectively works as a matching contribution, doubling the foundation's annual giving. "Buffett's gift came with three conditions for the Gates Foundation: Bill or Melinda Gates must be alive and active in its administration; it must continue to qualify as a

Bill and Melinda Gates with Warren Buffet in June of 2006.

charity; and each year it must give away an amount equal to the previous year's Berkshire gift, plus another 5% of net assets. Buffett gave the foundation two years to abide by the third requirement."[44] The Gates Foundation received 5% (500,000) of the shares in July 2006 and will receive 5% of the remaining earmarked shares in the July of each following year (475,000 in 2007, 451,250 in 2008, and so on).[45]

**Activities of the Foundation**

In the past, the foundation's grants have provided funds for such things as the prevention of AIDS and other diseases, and minority college scholarships. One such disease the Foundation is looking to eradicate is polio. B&MGF currently provides 17% (US$86 million in 2006) of the world budget for the attempted eradication. Mr. Gates and his wife have also donated over US$100 million to help children suffering from AIDS. Some of the other diseases that they are working on fighting are yellow fever, diphtheria, whooping cough and measles. In order to combat diseases such as these, Mr. Gates and his wife have contributed over US$750 million to the International Vaccine Fund.[46]

B&MGF has quickly become a major influence upon global health. The approximately US$800 million that the foundation gives every year for global health approaches the annual budget of the WHO and is comparable to the funds given to fight infectious disease by the USAID.[47] Below is a more detailed look at just of some the grants awarded by B&MGF. They include:

**Global Health**[48]

- US$750 million to The Global Alliance for Vaccines and Immunization.
- US$287 million to various HIV/AIDS researchers, which was split between 16 different research teams across the world, on the condition that they share their findings with one another.
- US$30 million for the new Department of Global Health at the University of Washington in Seattle.
- US$27 million to the Program for Appropriate Technology in Health (PATH): Children's Vaccine Program, to help vaccinate against Japanese encephalitis.
- US$10 million to The Institute for One World Health to support

the organization's work on a drug for visceral leishmaniasis (VL).

**Education**

- ❑ US$1 billion to the Gates Millennium Scholars which administers minority student scholarships via the United Negro College Fund.
- ❑ US$210 million to help outstanding graduate students outside of the U.K. study at the University of Cambridge.
- ❑ US$20 million for the University Scholars Program to endow a scholarship at Melinda Gates' alma mater, Duke University.

**In-depth look—Gates' global health vision**

In December of 2003, Tom Paulson, a reporter from the *Seattle Post-Intelligencer,* published several great articles on the B&MGF. He writes, "The battleground is disease prevention, not treatment. The tactics are businesslike performance, not traditional charitable giving ... The goal: break the developing world's cycle of sickness and poverty by eliminating the diseases that kill millions of children each year."[49]

The foundation so far has spent US$3.2 billion on hundreds of public health programs in more than 100 developing countries. Gates is especially interested in expanding access to basic and new vaccines. He is focused largely on preventing the spread of disease. The major benefit of the foundation is that Gates is demanding businesslike performance in an arena that in the past was basically good intentions poorly executed.

To implement Gates' strategy of disease prevention, GAVI, the Geneva-based vaccine initiative, was set up. What it does is it requires poor countries to submit a plan and apply for the money needed to buy vaccines. Gates is convinced that long-term charity doesn't work so B&MGF gives only to countries that can demonstrate progress against disease and are willing to invest in their own public health systems.[50]

Moreover, Gates believes that "Better health leads to economic improvements." This, in turn, will help better economies allow for more individual empowerment, which again in turn creates political demand for improvements in education, social infrastructure and on many other fronts. It is an interesting theory and one that I feel will work. "We have the resources and visibility to see if we can do something about these inequities," Gates said. "We can't do it on our own. But we might be able to catalyze the changes needed to break the cycle of disease and poverty."[51]

**Bono**

"As a rock star, I have two instincts, I want to have fun, and I want to change the world. I have a chance to do both."—Bono, U2 lead singer

A person very different from that of Gates and Winfrey is Paul Hewson. More commonly known as 'Bono,' he is the lead signer of the Irish band U2. He was named *Time Magazine* 2005 person of the year along with Bill

Bono talking to IMF Managing Director Horst Kohler (later Chancellor of Germany) and discussing debt relief at the Prague Congress Center in 2000.

Gates and his wife Melinda. Not too bad for a person who had the unholy and ironic nickname 'The Antichrist' from both family and friends when he was younger.[52] What most people are not aware of is that at the age of 15 Bono suffered a tragic and devastating loss when his mother died of a brain aneurysm while attending the funeral of her own father. Many people feel that it was this event that may be the reason Bono has such a restless and inquisitive nature.[53]

U.S. President George W. Bush described Bono as a, "doer ... The thing about this good citizen of the world is he's used his position to get things done."[54] His work began in 1984, when Bono appeared in Band Aid, which was a British and Irish charity group that was founded in 1984 by Bob Geldof and Midge Ure in order to raise money for famine relief in Ethiopia by releasing the record 'Do They Know It's Christmas?' for the Christmas market.[55] Bono also reprised his role in the 2004 Band Aid 20 and performed at Live Aid in 1985. In 2005, Bono played a huge role for Live 8, by organizing and publicizing the event, which was a series of ten concerts around the globe aimed at encouraging the representatives of the world's industrialized countries at the Group of Eight Summit to write off Africa's enormous debt, reform trade policy, and grant a great deal more aid for crises such as the AIDS epidemic.[56]

Since 1999, he has become increasingly involved in campaigning for third world debt relief and the plight of Africa. Furthermore, in March of 2002, Bono traveled to the White House for a special private meeting with President George W. Bush, who had just unveiled a US$5 billion aid package for the world's poorest countries that respect human rights. He also accompanied the President for a speech on the White House lawn. "This is an important first step and a serious and impressive new level of commitment ... This must happen urgently, because this is a crisis."[57]

Also in that year, Bono set up an organization called 'DATA', which stands for Debt, AIDS, Trade in Africa. The focus of the organization is to raise awareness about what he claims are Africa's unpayable debts, uncontrolled spread of AIDS, and unfair trade rules that hurt the continent's poor citizens. On the organization's website it states, "We talk to the experts, so we know what is really working—and what's not. We pull together, summarize and explain cutting-edge research on what works in Africa—and use our access to deliver those insights to top officials who might otherwise not hear the message of hope. And above all, we work to tell our leaders and politicians that people like YOU want to see action."[58]

In the spring of 2005 Bono, Ali Hewson and New York-based Irish fashion designer Rogan Gregory launched the socially conscious line EDUN. This has been an attempt to shift the focus in Africa from aid to trade. The goal here is by utilizing factories in Africa, South America, and India which provide fair wages to the workers and practice good business ethics EDUN will hopefully create a business model that will encourage others to invest in developing nations.[59]

In order to truly see the difference Bono is making, we must look at the 'Humanitarian Youth Culture Study' performed by Label Networks. They did a study of 13-24-year-olds in North America and U2's Bono was clearly the #1 Humanitarian Hero, as well as Angelina Jolie in third.[60] What is also interesting in the results are the high percentages for animated characters as Humanitarian or Environmental Heroes, such as Captain Planet and Smokey the Bear. While it may seem rather funny, it indicates potential in animated branding and marketing strategies that do good, e.g., a 'Sponge Bob' recycling program, or an anime AIDS awareness campaign.[61]

This study was the first report ever to quantify and provide in-depth 'green' marketing intelligence about the perceptions of 13-24-year-olds about the environment, humanitarian causes, volunteerism, and their future. As results clearly illustrate in the study, the timing is now to tap into these perceptions and trends to make a difference, not only in the North American youth culture landscape, but also towards effecting markets globally.[62]

Some of the results of this important study:

Name the greatest environmental or humanitarian hero:

- ❑ 16.6% Bono; 7.8% Gandhi; 6.5% Angelina Jolie; 5.8% Mother Theresa and David Suzuki; 4.4% Al Gore and Captain Planet

When it comes to humanitarian problems, what concerns you the most?

- ❑ 22.3% Poverty; 15.9% War; 14.2% AIDS; 13.2% Lack of Education; 10.5% Teen Suicide

Do you think about poverty in America or other parts of the world?

- ❑ 42.2% Yes, I think about it; 30.7% Yes I think about it sometimes; 24.1% Yes I think about it all the time and I want to do something about it

**Former Presidents Bill Clinton and Jimmy Carter**

I felt it would have been inappropriate to not include these two former presidents of the United States. Since the end of their presidencies both have been true humanitarians to the world.

## Clinton

"It is our duty as children of the same God and citizens of the same planet to pool our energies and banish the scourge of AIDS from the headlines of our newspapers to the chapters of our history books once and for all."—William J. Clinton

Since 2000, when his second term ended, Bill Clinton has been involved in numerous issues. Most of his more visible roles began at the start of 2005. Following the 2004 Indian Ocean earthquake, President George W. Bush named Clinton and George H. W. Bush to lead a nationwide campaign to help the victims of the disaster. Then on February 1, 2005, he was selected by then UN Secretary-General Kofi Annan to head the UN earthquake and tsunami relief and reconstruction effort. Then just five days later, he appeared with Bush on the Super Bowl XXXIX pre-game show in support of their bipartisan effort to raise money for relief of the disaster through the USA Freedom Corps. Two weeks later, they traveled to the affected areas to see the relief efforts.[63]

Later in the year, following Hurricane Katrina, Clinton again teamed with George H. W. Bush to coordinate private relief donations, in a campaign similar to their earlier one in response to the Indian Ocean tsunami.[64] Clinton was highly critical of the federal government's response to the hurricane, saying that the government "failed" the people affected, and that an investigation into the response was warranted.[65] Clinton has also been outspoken on the Bush administration's handling of climate change, publicly criticizing them at the 2005 UN Climate Change Conference.

## William J. Clinton Foundation

"The Clinton Global Initiative is making a difference by matching good ideas with the people who can implement them." William J. Clinton

© Clinton Foundation

Another major part of Clinton's post-presidency life has been his William J. Clinton Foundation. His vision is the foundation's mission: to strengthen the capacity of people in the U.S. and throughout the world to meet the challenges of global interdependence. Together with the generosity of citizens and volunteers, President Clinton and his foundation are vigorously working to advance those principles that move us beyond differences to a common future of shared responsibility, shared benefits, and shared values. Some of the foundation's programmes include: HIV/AIDS Initiative; Racial, Religious,

Ethnic Reconciliation Program; Economic Empowerment Program; and the Leadership Development and Citizen Service Program. While attending a Global Business Forum in Sydney, he signed a memorandum of understanding on behalf of his presidential foundation with the Australian government to promote HIV/AIDS programs in the Asia-Pacific region.[66]

Another program of his is the Clinton Global Initiative (CGI). According to the CGI website, it is a "non-partisan catalyst for action, bringing together global leaders including heads of state, non-profit organizations, charities, and business leaders to discuss challenges facing the world today and devise and implement innovative solutions to some of the world's most pressing challenges."[67] Clinton holds annual conferences, at which each invited guest must make a specific commitment to address one of the focus areas discussed.

Joe Conason of *Esquire Magazine* recalls the inaugural meeting in September 2005. In an era when celebrity dominates culture and consciousness, highly publicized events like the Clinton Global Initiative inevitably acquire the sheen of showbiz. Angelina Jolie, Brad Pitt, Mick Jagger, Barbra Streisand, Tony Bennett and Leonardo DiCaprio, were just a few of the names who attended.

Conason writes: "Clinton is brandishing a wireless microphone, obviously comfortable in the role of host and impresario. Soon he is to be joined onstage by Tony Blair, Condoleezza Rice, and the king of Jordan for a rambling and agreeable chat. Prompted by Clinton, they will talk about the Mideast peace process, terrorism, international trade, alternative energy and nuclear proliferation, sounding like guests on a high-minded chat show. Lurking not so inconspicuously in the reserved seats are current and former heads of state as well as rock legends and movie stars, and waiting for their turn to palaver with the former president are the chief executives of several of the world's largest corporations. Of course, this looks exactly like what his die-hard critics assumed the Clinton Global Initiative would turn out to be: just another elaborate occasion for him to seize the spotlight and shine it on himself."[68]

As Conason explains, this gathering appears to be nothing more than a Hollywood party, but looks can be deceiving. In the end, this 'Hollywood' party produced pledges of almost US$2 billion dollars to finance more than 200 new projects around the world. Some of the topics included climate change, global poverty, good governance and religious conflict and reconciliation.[69]

### Carter

"Globalization, as defined by rich people like us, is a very nice thing ... you are talking about the Internet, you are talking about cell phones, you are talking about computers. This doesn't affect two-thirds of the people of the world."—Jimmy Carter

One of the most successful organizations to date has been the Carter Center.

It was set up after his presidency with the help of his wife Rosalyn. Over the decades, The Carter Center's neutrality and record of achievement has given its peace-making programs the credibility needed to work nationally, regionally, and globally to advance peace and security. Moreover, the Center's work as an international peace organization improves lives worldwide.[70] You could write an entire book on just the Centers' accomplishments alone. For example, President Carter and The Carter Center have engaged in conflict mediation in Ethiopia and Eritrea (1989), North Korea (1994), Liberia (1994), Haiti (1994), Bosnia (1994), Sudan (1995), the Great Lakes region of Africa (1995-96), Sudan and Uganda (1999), and Venezuela (2002-2003). Under his leadership, The Carter Center has sent 61 international election-monitoring delegations to elections in the Americas, Africa, and Asia. These include Panama (1989), Nicaragua (1990), Guyana (1992), China (1997), Nigeria (1998), Indonesia (1999), East Timor (1999), Mexico (2000), Guatemala (2003), Venezuela (2004), Ethiopia (2005), and Liberia (2005).[71]

Carter is also a big promoter of global health. Some of the Center's accomplishments include:

- ❑ As Guinea worm is poised to become the next disease after smallpox to be wiped off the face of the Earth, eradication efforts are rejuvenating communities throughout Africa, enabling children to return to school and farmers to their fields. The Center spearheads the international campaign, which has reduced cases by more than 99.5% since 1986.
- ❑ Improved sanitation and hygiene is critical to public health and overall development. As part of the Center's trachoma control efforts, more than 200,000 latrines have been built in Ethiopia since 2004, transforming life in those communities.
- ❑ They have assisted in the delivery of more than 75 million treatments in 11 river blindness-endemic countries in Latin America and Africa since 1996. The Carter Center is leading the drive to eradicate this debilitating disease where it occurs in the Americas by the end of the decade.
- ❑ Building on village-based drug distribution systems now in place in Nigeria to prevent river blindness, the Center also is reducing incidences of the parasitic disease schistosomiasis, enhancing children's abilities to grow, develop, and learn.
- ❑ The same healthcare delivery infrastructure in Nigeria also is helping to prevent and treat lymphatic filariasis, a disfiguring and shame-ridden disease afflicting the poorest of the poor.
- ❑ In some countries, the biggest factor to poor health is lack of access to trained health personnel. By strengthening the training of Ethiopia public health staff, 90% of the people in this large African country, who live in less developed rural areas, have greater access to basic disease prevention and healthcare services.

Local health promoters visit communities to distribute Mectizan® drug treatments (donated by Merck) every six months. Health education and Mectizan are key elements in the fight to eliminate river blindness, a preventable disease caused by the bite of a small, black fly.

- More than 4 million small-scale farmers in 15 Sub-Saharan African countries have learned improved agricultural techniques to double or triple grain production, growing more food for their families and boosting local economies.
- Because mental and physical health are interconnected, the Center has led international efforts to reduce stigma and discrimination against people with mental illnesses and to achieve greater equity for mental health in the healthcare system.
- The Center is the base for the International Task Force for Disease Eradication. The group has reviewed more than 100 infectious diseases and identified six as potentially eradicable.[72]

Carter has also been involved in several other programs. For instance, since 1984, he has worked with Habitat for Humanity International (HFHI). During that first year, the former president led a work group to New York City to help renovate a six-storey building with 19 families in need of decent, affordable shelter. "That experience planted the seed, and the Jimmy Carter Work Project has been an internationally recognized event of HFHI ever since."[73]

Besides all of this, since his presidency, Carter has become a prolific author writing on a variety of topics including his presidency, the Middle East, and his Christian faith. He even wrote a collection of poetry.[74]

All in all, I think Bill Clinton described Carter the best. "I've never met anyone who did more with the gifts God gave him than Jimmy Carter," Clinton said, holding up the unlit stub of a thick cigar in his right hand while fanning the cards in his left. "He declared what he was interested in and where he thought he could make a difference. They were superficially disparate, but all helped people have a better life. He eradicated river blindness in Africa, he worked on the green revolution to help people become more agriculturally self-sufficient, he's advanced human rights and monitored elections ... He has things that he plans to do, and things that come up."[75]

## Lesser-known people

After reading about these particular individuals, that same question from the beginning of the chapter might be resurfacing. What difference can I make? Like I said earlier, we can all make a difference and that is something we must do. To further prove to people who are still questioning themselves, let's look at a few lesser known people who have made a substantial impact on the world.

### Millard and Linda Fuller

"I see life as both a gift and a responsibility. My responsibility is to use what God has given me to help his people in need."—Millard Fuller

Millard Fuller studies house plans with a man in Zaire, circa 1974.

In 1942, a Christian farming community was founded just outside of Americus, Georgia in the United States. Koinonia Farm was its name, and it was a small, interracial community founded by farmer and biblical scholar Clarence Jordan. In 1965, the Fullers first visited Koinonia. They had recently left a successful business in Montgomery, Alabama (USA) to begin a new life of Christian service. What spawned out of this little community was the concept of "partnership housing" which was developed by Jordan and Fuller.[76]

The concept was simple: those in need of adequate shelter would work side by side with volunteers to build simple, decent houses. The houses would be built with no profit added and no interest charged. Building would be financed by a revolving Fund for Humanity. The fund's money would come from the new homeowners' house payments, donations and interest-free loans provided by supporters and money earned by fund-raising activities.[77] The monies in the Fund for Humanity would be used to build more houses.

In 1968, Koinonia laid out 42 half-acre house sites with four acres reserved as a community park and recreational area. Money was donated from around the country to start the work. Homes were built and sold to families in need at no profit and no interest, thus the basic model of Habitat for Humanity was begun. It spread like wildfire, and in 1973, the Fullers decided to apply the Fund for Humanity concept in developing countries, going to Zaire (now the DRC) to build 2,000 houses. Three years later, they returned to the U.S.

and not soon after, Habitat for Humanity International was created.[78]

Through the work of Habitat for Humanity, thousands of low-income families have found new hope in the form of affordable housing. Due to the Millard's vision and their hard work and dedication, churches, community groups and others have joined together to successfully tackle a significant social problem—decent housing for all. Today, Habitat for Humanity has built more than 200,000 houses, sheltering more than 1,000,000 people in more than 3,000 communities worldwide.[79]

Habitat for Humanity dedicated its 200,000th house in 2005. It is now the home of the Kouassi-Harper family of Knoxville, Tenn. (USA).

**Jody Williams**

As with Millard and Linda Fuller, not too many people are familiar with Jody Williams. She started her career by working for eleven years to build public awareness about the U.S. government policy toward Central America. From 1986 to 1992, she developed and directed humanitarian relief projects as the deputy director of the Los Angeles-based Medical Aid for El Salvador. From 1984 to 1986, she was co-coordinator of the Nicaragua-Honduras Education Project, leading fact-finding delegations to the region. Also in those first eleven years, Jody taught English as a Second Language (ESL) in Mexico, the U.K., and Washington, D.C.[80]

Jody Williams giving a speech in Japan, 1999.

Then in October of 1992, as the founding coordinator, she helped launch the International Campaign to Ban Landmines (ICBL). On that day, the ICBL began with only six NGOs but as its coordinator, Williams oversaw the ICBL grow to more than 1,300 NGOs in over 85 countries. Serving as both a strategist and spokesperson for the campaign, she achieved the ICBL's goal in September 1997, when an international treaty banning antipersonnel landmines was signed during a diplomatic conference in Oslo, Norway.

Jody has continued her work, speaking on topics such as human security, human rights and international law, on the role of civil society in international diplomacy, and on individual initiative in bringing about social change. In addition, from 1999-2004, she served as senior editor for its 1,000-page annual Landmine Monitor Report, a groundbreaking initiative by civil society that monitors the implementation and compliance of the Mine Ban Treaty.[81]

**Clinton Hill**

"Clinton Hill was special not because cancer tragically took his life at the age of eleven, but because the life he did live displayed a courage and sensitivity to the world around him."[82] During his short time on this planet, Clint just didn't understand what we are doing to our planet. Why would we dump poisons into our rivers and oceans? Why would we neglect our precious plants and animals? Many of us still wonder that today.

But this is why Clint was different; he didn't just wonder, he acted. Clint started a club for kids dedicated to peaceful Earth-saving actions. He called it Kids for Saving Earth. After Clint died, his mother and father established Kids for Saving Earth(KSE) as a non-profit organization with thousands of kids doing good Earthworks. In the past, KSE kids have given speeches at the UN, rode on a KSE float in President Bill Clinton's inaugural parade and participated in thousands of grassroots activities to help protect the environment. Today, KSE is a network of over 7,000 schools and environmentally concerned kids and adults.[83]

The main mission of KSE is to educate, inspire, and empower children to protect the Earth's environment. KSE Worldwide provides action-oriented educational posters and information support materials to kids, families, groups, classrooms and schools. A few examples of programs KSE offers include: The KSE Children's Forest, The Arctic is Cool, The KSE Rainforest, The Rock the World Concert Kit and CD, The Call to Action, The Wonderful World of Water and much more.[84]

## Nobel Peace Prize

Although a prize cannot compare to helping humanity and the planet Earth, the Nobel Peace Prize does have the ability to turn the world's gaze onto some problem or other, one that might go unnoticed were it not for the prize. However, the real benefit of the Nobel Peace Prize is that it shows one person can make a difference. For those who are not familiar with the Nobel Peace Prize, it is awarded annually to a certain

Alfred Nobel.

individual or group who has made a significant contribution to humanity. There are over 300 peace prizes in the world but none as well known and as highly respected as the Nobel Peace Prize.

The history of the award dates back to the 19th century. When Alfred Nobel died on December 10, 1896, it was discovered that he had left a will, dated November 27, 1895, according to which most of his vast wealth was to be used for five prizes, including one for peace. But how do you get awarded the Nobel Peace Prize? One important element of this record has been the committee's broad definition of peace, enough to take in virtually any relevant field of peace work. Since the annual prize was first awarded in 1901, the Nobel Foundation has recognized individuals and organizations for their efforts in such fields as humanitarian work, peace movements, human rights efforts, mediation of international conflicts, arms control and disarmament.[85]

The individuals who are willing to go the extra mile are always the frontrunners in winning the Nobel Peace Prize. Some of the winners (mentioned throughout the chapter) include Nelson Mandela (1993) and Jimmy Carter, who was awarded the peace prize in 2002. Winfrey, Jolie, and Bono have all traveled the world to raise AIDS awareness, and Bono is almost a one-man crusade for Third World debt forgiveness. Bono was nominated for the prize in 1995 but didn't make the final cut.[86] On December 10, 2006, it was announced that the winner was economist Muhammad Yunus from Bangladesh along with his Grameen Bank. For more than two decades they have helped poor people start their own businesses by providing small, usually unsecured loans known as microcredit.

In 1997, Jody Williams was recognized, along with the ICBL, with the Nobel Peace Prize. In one particular interview, Jody was asked if she ever thought she would win the Nobel Peace Prize. She answered, "We had no idea where the campaign would lead, but we knew we could do something to make the lives of mine victims better and that was what motivated us then and motivates us today. The real 'prize' for us in the campaign is the Mine Ban Treaty—which provides the world a framework to ultimately eliminate this indiscriminate killer of men, women and children."[87]

Her answer to that question is why Jody Williams is such a great role model. It is not about the prize, but it is about making the world a better and safer place.

## Conclusion

As you have just read, individuals all over the world are continuing to make a difference. It doesn't matter how famous you are, or how much money you have, one can make a substantial impact.

People need to know that they can make a difference. People of all ages, but especially the younger generations, need to be encouraged to recognize this and act upon it. Sometimes you might feel that it is easier to keep quiet and let things go on as they are, but hopefully you have learned from the individuals in this chapter that you have to stand up and fight.

The concept of one person changing the course of history is an essential lesson that needs to be learnt from this chapter.

In 1995, during an interview conducted by Dawn Engle and Ivan Suvanjieff at the residence of His Holiness, the 14th Dalai Lama (also a Nobel Peace Prize winner, 1989), in Dharamsala, India, they asked him two very important questions. The first question was, "You are a great man, and I am a very small person. How can I make any sort of difference in this world?" He responded by saying, "I do not consider myself to be something special. I always consider all human beings to be special, whether they are educated or uneducated, they all have the human mind and the human spirit and that is what is important. So now the question is whether you utilize that human potential with confidence. That is the question. Every human being has the same potential without losing confidence, with great determination. You need to try."

The second question was, "How much time do we have left on planet earth, given the environmental destruction and all the damage we have done. Is there hope for survival on this planet?" He replied, "In spite of many problems, much suffering, many tragic situations, I believe the human race has the ability to survive and not only to survive, but also I think because of our experiences over the last century we have learned so much. Perhaps as a result of our experiences in these complicated, difficult times, I think humanity in general is becoming more mature. So if we really put every effort, more concentrated effort, with more clear vision about the future, I think our next century will certainly be a much happier one. The future is on our own shoulders. If we find failure—again make an effort. Failure—again make an effort. There is always the possibility to have a better future. If we give up hope and we give up the effort, then we will be asking for ruin and destruction."[88]

# Chapter 7

# The contribution of international organizations

"In this globalized world, fewer and fewer problems affect only individual countries, and they certainly cannot be solved by individual countries' actions."—Wolfgang Thierse, German politician

The United Nations headquarters in New York City, USA. It is home to the General Assembly, Security Council, the Secretariat and other departments and agencies.

In the previous chapter we read about individuals who have dedicated themselves to combating many of the global issues. In addition to those individuals there are thousands of people, from all corners of the globe, who work for various international organizations that are also committed to that same goal.

Since the end of World War II, various international organizations have been created both to maintain and to help create world peace. For many, this is the correct path to peace. In the world of international politics, organizations are generally broken down into two categories: intergovernmental organizations (IGOs) and nongovernmental organizations (NGOs). Both have existed prior to the 20th century and both continue to rise in numbers.

## Intergovernmental organizations

IGOs have been created by states to solve their shared problems.[1] IGOs' members are states themselves and thus are seen as more important than

NGOs. The IGOs get their power and authority from the states that assign them.

Kegley and Wittkopf state, "IGOs are defined not only by the fact that their members are the governments of states, but also by their performance and institutional organization; IGOs meet at relatively regular intervals and have specified procedures for making decisions and a permanent secretariat or headquarters staff."[2] IGOs also vary in their size and purpose. Some are intercontinental organizations like the UN and others may focus on one particular topic or region. For example, NATO or the North Atlantic Treaty Organization, is a military alliance mostly dealing with Europe.

As you can imagine, many of these organizations overlap each other as they deal with a broad range of global issues. When this happens, they support one another. Some of these issues include, "trade, defense, disarmament, economic development, agriculture, health, culture, human rights, the arts, illegal drugs, tourism, labor, women's plight, education, debt, the environment, crime, humanitarian aid, civilian crisis relief, telecommunications, science, globalization, immigration, and refugees."[3] For a simple classification, see Kegley and Wittkopf's table below.

**A simple classification of IGOs[4]**
Range of Stated Purpose

| Geographic scope of membership | Multiple purposes | Single purpose |
|---|---|---|
| **Global** | United Nations World Trade Organization | World Health Organization |
| | UNESCO | International Labour Organization |
| | Organization of the Islamic Conference | International Monetary Fund |
| | | Universal Postal Union |
| **Inter-regional, regional and sub-regional** | African Union | European Space Agency |
| | European Union | Nordic Council |
| | Organization for security and cooperation in Europe | North Atlantic Treaty Organization |
| | Organization of American States | International Olive Oil Council |
| | League of Arab States | African Groundnut Council |

**United Nations**

The UN is a huge organization that covers many different fields. Due to its size it is not possible for me to provide a true in-depth analysis of the entire organization. If you do want to read more, I encourage you to visit the UN's website (www.un.org).

Many are familiar with the UN but are not entirely sure what they do or how they are set up. Currently there are 191 'Member States' from every region of the world. These 191 member states make up what is called the General Assembly or the 'GA'. This 'parliament of nations', which meets regularly (September-December), and in special sessions, comes together to address various global problems. Each Member State has one vote.

There is also what is called the UN Security Council (UNSC). Their primary responsibility is to maintain international peace and security. The Council may convene at any time, whenever peace is threatened. Under the UN Charter, all member states are obligated to carry out the Council's decisions, whatever they may be. There are 15 Council members in total of which five are permanent members. They are China, France, Russia, the U.S. and the U.K.. The other ten are elected by the GA for two-year terms.

Decisions of the Council require nine yes votes. Except in votes on procedural questions, a decision cannot be taken if there is a no vote, or veto, by a permanent member.[5] This is known as the 'veto' power. When there is a threat to international peace, the UNSC first tries ways to settle the dispute peacefully. In the event of fighting, the Council tries to secure a ceasefire, such as in 2006 Israel/Lebanon conflict. It may also send a peacekeeping mission to help the parties maintain the truce and to keep opposing forces apart. In addition, the UNSC can take action by imposing economic sanctions or by ordering an arms embargo. On rare occasions, the Council has authorized member states to use "all necessary means", including collective military action, to see that its decisions are carried out.[6]

The head of the UN is known as the Secretary-General or SG. At the end of 2006, Kofi Annan of Ghana, who served for ten years, was replaced by South Korea's Foreign Minister Ban Ki-moon. He is the eighth SG and will serve a five-year term.

After his appointment by the GA in October 2006, Mr. Ban noted that the road towards peace, prosperity and dignity for all has many pitfalls. "As Secretary-General, I will make the most of the authority invested in my office by the Charter and the mandate you give me. I will work diligently to materialize our responsibility to protect the most vulnerable members of humanity and for the peaceful resolution of threats to international security and regional stability,"[7] he said. Mr. Ban has been praised for his diplomatic skills and personal qualities, and said the world body would be in able hands as it faces the challenges to come.[8]

Switching topics, you may be wondering where does the UN get all its money. The UN budget has always been and will continue to be a big issue. It consists of three parts: the core budget, the peacekeeping budget, and the

UN Secretary General Ban Ki-moon (left) on one of his many trips to Asia.

budget for voluntary programs.[9] Regarding the peacekeeping and the voluntary programs, member states contribute as they see fit. With the core budget, each member contributes a certain amount of money depending on their capacity to pay. For instance, the U.S. which has the greatest resources contributes the most amount of money to the UN budget. Due to this setup, the richest states pay for more than three-fourths towards the UN's billion dollar budget.[10]

Going back to the end of World War II, the UN was established on 24 October 1945 by 51 countries committed to preserving peace through international cooperation and collective security.[11] When a state becomes a member of the UN, they agree to accept the UN Charter. The Charter is an international treaty that sets out basic principles of international relations. According to the Charter, the UN has four purposes: to maintain international peace and security; to develop friendly relations among nations; to cooperate in solving international problems and in promoting respect for human rights; and to be a centre for harmonizing the actions of nations.[12]

You must understand that the UN is not a world government and it does not make laws. It does, however, provide the means to help resolve international conflicts and formulate policies on matters affecting all of us.[13] The UN has six main organs. Five of them—the General Assembly, the Security Council, the Economic and Social Council, the Trusteeship Council and the Secretariat—are based at UN Headquarters in New York. The sixth, the International Court of Justice, is located at The Hague in the Netherlands.

The organization is extremely important in solving many global problems. To help do this they are more than 30 affiliated organizations, known collectively as the UN system. The UN website states, "Day in and day out, the UN and its family of organizations work to promote respect for human rights, protect the environment, fight disease and reduce poverty. UN agencies define the standards for safe and efficient air travel and help improve telecommunications and enhance consumer protection. The UN leads the international campaigns against drug trafficking and terrorism. Throughout the world, the UN and its agencies assist refugees, set up programmes to clear landmines, help expand food production and lead the fight against AIDS."[14]

Let's take a closer look at some of the specialized agencies. They are:

**FAO (Food and Agriculture Organization of the UN)** [15]
Works to improve agricultural productivity and food security, and to better the living standards of rural populations.
**IAEA (International Atomic Energy Agency)**
An autonomous intergovernmental organization under the aegis of the UN, it works for the safe and peaceful uses of atomic energy.
**ICAO (International Civil Aviation Organization)**
Sets international standards for the safety, security and efficiency of air transport, and serves as the coordinator for international cooperation in all areas of civil aviation.
**IFAD (International Fund for Agricultural Development)**
Mobilizes financial resources to raise food production and nutrition levels among the poor in developing countries.
**ILO (International Labour Organization)**
Formulates policies and programmes to improve working conditions and employment opportunities, and sets labour standards used by countries around the world.
**IMF (International Monetary Fund)**
Facilitates international monetary cooperation and financial stability and provides a permanent forum for consultation, advice and assistance on financial issues.
**IMO (International Maritime Organization)**
Works to improve international shipping procedures, raise standards in marine safety and reduce marine pollution by ships.
**ITU (International Telecommunication Union)**
Fosters international cooperation to improve telecommunications of all kinds, coordinates usage of radio and TV frequencies, promotes safety measures and conducts research.
**UNESCO (UN Educational, Scientific and Cultural Organization)**
Promotes education for all, cultural development, protection of the world's natural and cultural heritage, international cooperation in science, press freedom and communication.
**UNIDO (UN Industrial Development Organization)**
Promotes the industrial advancement of developing countries through technical assistance, advisory services and training.
**UPU (Universal Postal Union)**
Establishes international regulations for postal services, provides technical assistance and promotes cooperation in postal matters.
**WHO (World Health Organization)**
Coordinates programmes aimed at solving health problems and the attainment by all people of the highest possible level of health. It works in such areas as immunization, health education and the provision of essential drugs.
**WIPO (World Intellectual Property Organization)**
Promotes international protection of intellectual property and fosters

cooperation on copyrights, trademarks, industrial designs and patents.

**World Bank Group**

Provides loans and technical assistance to developing countries to reduce poverty and advance sustainable economic growth.

**WMO (World Meteorological Organization)**

Promotes scientific research on the Earth's atmosphere and on climate change, and facilitates the global exchange of meteorological data.

**WTO (World Tourism Organization)**

Serves as a global forum for tourism policy issues and a practical source of tourism know-how.

The UN has done some remarkable work in its 62-year existence. It has helped defuse international crises and resolve conflicts. It has undertaken complex operations involving peacemaking, peacekeeping and humanitarian assistance. Under the Charter, member states agree to settle disputes by peaceful means and refrain from threatening or using force against other states.[16]

According to the UN website, UN efforts have produced dramatic results. The UN helped defuse the Cuban missile crisis in 1962 and the Middle East crisis in 1973. In 1988, a UN-sponsored peace settlement ended the Iran-Iraq war, and the following year UN-sponsored negotiations led to the withdrawal of Soviet troops from Afghanistan. In the 1990s, the UN was instrumental in restoring sovereignty to Kuwait and played a major role in ending civil wars in Cambodia, El Salvador, Guatemala and Mozambique, and resolving or containing conflict in various other countries.[17]

### Millennium Development Goals

During the UN Millennium Summit held in September 2000, world leaders agreed to a set of time-based and measurable goals and targets for combating poverty, hunger, disease, illiteracy, environmental degradation and discrimination against women.[18] Now referred to as the Millennium Development Goals or MDGs, they provided a framework for the whole UN system to work together. The time limit was set for 2015 with the hope of achieving the following eight main goals:

- Halve extreme poverty and hunger;
- Achieve universal primary education;
- Empower women and promote equality between women and men;
- Reduce under-five mortality by two-thirds;
- Reduce maternal mortality by three-quarters;
- Reverse the spread of diseases, especially HIV/AIDS and malaria;
- Ensure environmental sustainability;
- Create a global partnership for development, with targets for aid, trade, and debt relief.[19]

Another big part of the MDGs is the monitoring aspect. UN agencies such as the World Bank, OECD and the IMF are assisting with MDG reports for every developing country. Recent news suggests that the implementation of the MDGs is not going so well. One of the countries that accepted the MDGs is India. The experts point out however, that the country is way off the mark because of the lacklustre policies of the government and the private sector.[20]

Muchkund Dubey, president of the council for social development states, "We need to understand and enforce the constitutional rights given to the poor which they are unaware of. We need to adopt programmes and policies which are targeted towards the needy and we should differentiate between poverty eradication and poverty reduction."[21]

Another large problem seems to be with the media. Discussing the role of the media in spreading awareness about the MDGs, senior journalist Usha Rai said: "Newspapers only look for business, whatever they write is about is urban-centric. I have met a large section of educated and learned journalists who have not even heard of the goals."[22] Furthermore, according to Rai, the goal of providing education under the Sarva Shiksha Abhiyan has not been taken seriously by the government and the media.

**United Nations peacekeeping**

Most are familiar with the famous 'blue helmets'. They are sent to flashpoints all over the world. In the first 40 years of their existence, they were only 13 UN peacekeeping missions. In the last 20 years, there have been over thirty-eight.[23]

A protective mask and grenade launcher are part of the uniform of this Nepalese soldier.

Since the end of the Cold War, UN peacekeepers have been called on to preserve international security. In many of their first few missions after the Cold War, their results were substandard, but understandably so. They have seemed to learn from their mistakes, but the fact of the matter is the UN is not yet a true collective security organization. In fact, peacekeeping is not even mentioned anywhere in the UN charter.

Even though the Department of Peacekeeping Operations (DPKO) has not been totally successful, they now face an even bigger problem than resolving conflicts—money. The UN is currently undergoing major reform. Moreover, the UN is in a budget crunch. At a time when it seems the peacekeepers may be needed the most, the UN will not be in a position to help everyone. This is known as selective engagement when the UN is forced to choose which conflicts or situations they want to intervene in.

As of February 2005, 103 countries were contributing nearly 67,000 uniformed personnel—a record number.[24] However, UN peacekeeping operations still remain underfunded, understaffed and overwhelmed.

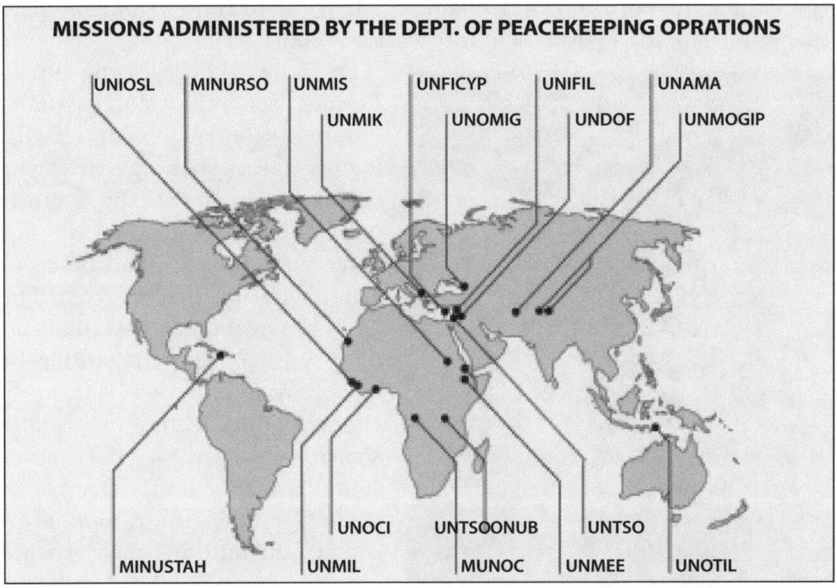


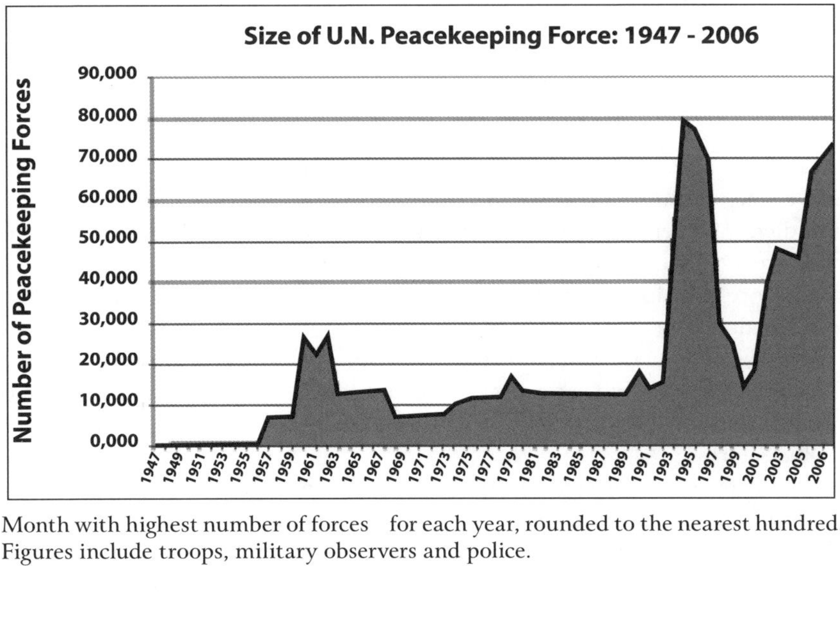


Month with highest number of forces for each year, rounded to the nearest hundred. Figures include troops, military observers and police.

**International criminal court**

The idea for an international court to deal with the problem of crimes committed against humanity really started after the Nuremburg Trials which dealt with the war crimes committed by the Nazis in World War II. Not much came of the idea until the conflicts in Rwanda and Yugoslavia. Following both events, *ad hoc* tribunals were established: the International Criminal Tribunal for Rwanda in 1994, and the International Criminal Tribunal for the Former Yugoslavia in 1993.[25] Due to those tribunals there became a desire for a permanent international court.

The ICC headquarters in The Hague, The Netherlands.

On July 17, 1998 the Rome Statute of the International Criminal Court was adopted with almost all states participating voting in favor of the statute. Nevertheless, some states did not agree with the idea and failed to sign or ratify the treaty.[26] The U.S. is one of those states that have not been quiet regarding its disapproval of the ICC. Strangely enough Bill Clinton initially signed the treaty, but never submitted it for ratification. Then when George W. Bush took office in 2001, he nullified the signature with support of both Republicans and Democrats.[27] Nonetheless, in 2002 the Rome Statute became a binding treaty when 60 states ratified it and with that, the ICC legally came into existence. Today there are over 100 states who are party to the ICC.[28]

The ICC's main goal is to prosecute individuals for war crimes, genocide and crimes against humanity. In addition, the Court can only exercise its jurisdiction when national courts are unwilling or unable to investigate or prosecute such crimes, thus being a 'court of last resort".[29] Therefore, primary responsibility to exercise jurisdiction over alleged criminals is left to individual states.

The ICC has a relationship with the UN via a relationship agreement. This agreement provides for the UNSC to provide referrals under the Rome Statute, and for UN assistance in payment for any prosecutions made under such a referral.

Countries who ratify the treaty that created the ICC grant it authority to try their citizens for war crimes, crimes against humanity, and genocide.[30] This has become a very controversial issue. Furthermore, many states want to add 'aggression', 'terrorism' and 'drug trafficking' to the list of crimes covered by the Rome Statute. Other states are opposed to this because these crimes are difficult to define, and that dealing with less serious crimes such as terrorism and drug trafficking would distract from the seriousness of the crimes for which the ICC was established.[31] As a compromise, the treaty includes the word 'aggression' as a crime, but does not provide a definition. This definition is pending adoption of an amendment to the statute. It may also be amended to include other crimes, however, no amendments can be made to the Rome Statute until seven years after the statute became legally binding.[32]

In its short history, the ICC has seen three important cases. Wikipedia highlights below:

- Uganda and the Lord's Resistance Army (LRA), which was referred to the court on January 29, 2004 by Republic of Uganda, a 'state party' of the court (the term indicates that they are a country that has ratified the court's Rome Statute). On October 6, 2005 the court issued its first public arrest warrants for the Lord's Resistance Army leader Joseph Kony, his deputy Vincent Otti, and LRA commanders Raska Lukwiya, Okot Odiambo, and Dominic Ongwen.
- The situation in Ituri, Democratic Republic of Congo (DRC), which was referred to the court by the DRC on April 19, 2004. On March 17, 2006 Thomas Lubanga, former leader of the Union of Congolese Patriots militia in Ituri, became the first person to be arrested under a warrant issued by the court.
- The situation in Darfur, Sudan, which was referred to the court in March 2005 by the UNSC.

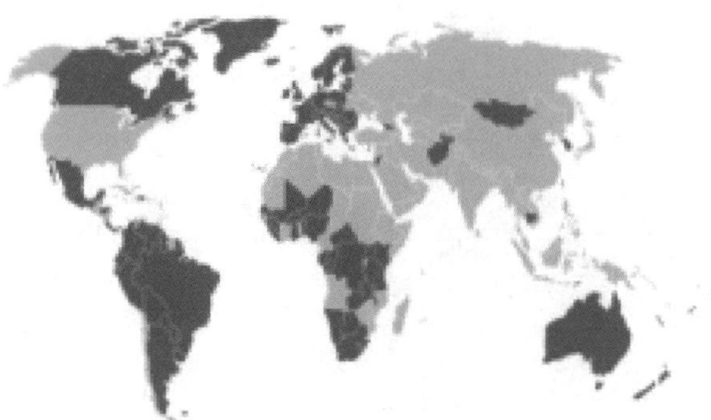

World map of ICC member states. As of August 2006, 102 countries have ratified or acceded to the court, including nearly all of Europe and South America, and nearly half of all African countries. A further 39 other states have signed but not yet ratified the treaty, which under customary international law obliges a state to refrain from acts which would defeat the object and purpose of the treaty. The U.S. and Israel have 'unsigned' the Rome treaty in order to avoid these obligations.[34] (Source: Wikipedia)

Although the ICC has support from many states, others do not share the same optimism as they do. American Tom DeLay, describes the ICC, which is backed by nearly 100 countries, including Canada, and all of Washington's traditional European allies except Turkey, as UN Secretary-General "Kofi Annan's Kangaroo Court" whose legitimacy was "laughable."[3]

**Interpol**

The world's largest international police organization with 184 member countries, the International Criminal Police Organization or Interpol, was created in 1923. It facilitates cross-border police cooperation, and supports and assists all organizations, authorities and services whose mission is to prevent or combat international crime. Even when diplomatic relations do not exist between particular countries, Interpol aims to facilitate international police cooperation.[35] According to their website, "Action is taken within the limits of existing laws in different countries and in the spirit of the Universal Declaration of Human Rights. Interpol's constitution prohibits any intervention or activities of a political, military, religious or racial character."[36]

Interpol has a General Secretariat located in Lyon, France, which operates 24 hours a day, 365 days a year and is run by its own Secretary-General. Officials from more than 80 countries work side by side in any of the organization's four official languages: Arabic, English, French and Spanish. The Secretariat has six regional offices: in Argentina, Côte d'Ivoire, El Salvador, Kenya, Thailand and Zimbabwe, and a liaison office at the UN in New York.[37] In addition to the General Secretariat in Maand and the regional offices, each Interpol member country maintains a National Central Bureau (NCB) staffed by national law enforcement officers. The NCB is the designated contact point for the General Secretariat, regional offices and other member countries requiring assistance with overseas investigations and the location and apprehension of fugitives.[38]

Lyon, France, March 1-2, 2005. Interpol Secretary-General Ronald K. Noble welcomes more than 500 delegates from 155 countries to the 1st Interpol Global Conference on Preventing Bio-terrorism.

One of Interpol's most important functions is to help police in member countries share critical crime-related information using the organization's system of international notices.[39] The notices are put out concerning individuals wanted for serious crimes, missing persons, unidentified bodies, possible threats and criminals' modus operandi. Additionally, notices are used by the UN to alert police of individuals subject to UN sanctions against Al Qaeda and the Taliban, and by the

International Criminal Tribunals for the Former Yugoslavia and Rwanda to seek persons wanted for serious violations of international human rights laws.[40]

Interpol's notices contain various information such as identity details, physical description, photographs, fingerprints and other relevant information such as occupation and languages spoken. If the person has been charged for a crime, the notice will most likely contain all his or her judicial information like the offence the person was charged with, what laws were broken, the maximum penalty and so on. Listed here are the notice descriptions with their corresponding color.

**Red Notice**: To seek the arrest or provisional arrest of a wanted person with a view to extradition based on an arrest warrant.
**Blue Notice**: To collect additional information about a person's identity or illegal activities in relation to a criminal matter.
**Green Notice**: To provide warnings and criminal intelligence about persons who have committed criminal offences and are likely to repeat these crimes in other countries.
**Yellow Notice**: To help locate missing persons, especially minors, or to help identify persons who are not able to identify themselves.
**Black Notice**: To seek information about unidentified bodies.
**Interpol-UNSC Special Notice**: To alert police of groups and individuals who are the targets of UN sanctions against Al Qaeda and the Taliban.
**Orange Notice**: To warn police, public entities and other international organizations of possible threats from hidden weapons, parcel bombs and other dangerous materials.

All notices are available to authorized law enforcement personnel via Interpol's secure website. In addition, some Red and Yellow notices are published on the organization's public website with the agreement of the member country directly concerned.[41]

**Key figures**[42]

The General Secretariat issued 3,104 notices in 2005:

| Red | Blue | Green | Yellow | Black | Interpol-UN | Orange |
|---|---|---|---|---|---|---|
| 2 206 | 237 | 304 | 272 | 74 | 4 | 7 |

**Regional organizations**

In recent years, dozens of regional trade agreements have been registered all over the world. Two of the better known regional trade organizations are the North American Free Trade Association (NAFTA) and the Asia-Pacific Economic Cooperation (APEC). There has also been a growth in regional security organizations. NATO is probably the best known, but others exist

The OSCE, UN, EU, CoE, NATO and other international organizations are working to find solutions for a broad range of security issues and challenges.

like the Organization for Security and Cooperation in Europe (OSCE).

The UN also has been very willing to hand over responsibility for a variety of issues to the regional organizations. The UN itself reached this conclusion in a recent report, saying: "The United Nations does not have, at this point in its history, the institutional capacity to conduct military enforcement measures under Chapter VII (of the UN Charter). Under present conditions, ad hoc member states coalitions of the willing offer the most effective deterrent to aggression or to the escalation or spread of an ongoing conflict ... The Organization still lacks the capacity to implement rapidly and effectively decisions of the Security Council calling for the dispatch of peacekeeping operations in crisis situations. Troops for peacekeeping missions are in some cases not made available by member states or made available under conditions that constrain effective response. Peacemaking and human rights operations, as well as peacekeeping operations, also lack a secure financial footing, which has a serious impact on the viability of such operations."[43]

To get a better feel of the roles and responsibilities of some of the larger regional IGOs, let's take a closer look at the Organization of American States (OAS).

Dating back to 1889, the OAS is the world's oldest regional organization. Their charter was signed in Bogota, Colombia in 1948 and entered into force in December 1951. Today, the OAS has 35 member states. In addition, the Organization has granted Permanent Observer status to 37 States, as well as the EU.[44]

The basic purposes of the OAS are as follows: to strengthen the peace and security of the continent; to promote and consolidate representative democracy, with due respect for the principle of non-intervention; to prevent possible causes of difficulties and to ensure the peaceful settlement of disputes that may arise among the member states; to provide for common action on the part of those States in the event of aggression; to seek the solution of political, juridical and economic problems that may arise among them; to promote, by cooperative action, their economic, social and cultural development, and to achieve an effective limitation of conventional weapons that will make it possible to devote the largest number of resources to the economic and social development of the member states.[45]

In December 1994, the democratically elected heads of state and government of the American nations came together for the first time in Miami, Florida (USA). The leaders declared that "the strengthening, effective exercise and consolidation of democracy constitute the central political priority of the Americas. The Organization of American States (OAS) is the principal hemispheric body for the defense of democratic values and institutions."[46] They also noted that the OAS had a particularly important role to play in supporting the strengthening of democracy; the promotion and protection of human rights; the struggle against corruption; efforts to eliminate the threat of national and international terrorism; efforts to foster mutual confidence; free trade in the Americas; and telecommunications and information infrastructure. The Plan of Action of the Miami Summit also envisages other roles for the OAS in the following areas: promoting cultural values; combating the problem of illicit drugs and related crimes; cooperation in science and technology; strengthening the role of women in society; and partnership for pollution prevention.[47]

## Non-governmental organizations

Non-governmental organizations or NGOs greatly differ from IGOs. NGOs are private international actors whose members are not states, but instead drawn from the populations of two or more states who have formed organizations to promote their shared interests and ideals in order to influence the policies of state governments and IGOs.[48] There are over 30,000 NGOs all over the world.

Participants at a SaferAfrica training course prior to Operation Rachel XI, a weapons and ammunition destruction, crime-combating operation that has been jointly executed by the police services of South Africa and Mozambique for the last eleven years. This destruction of weapons has made a large impact on the availability of weapons on the illicit market, not just in South Africa and Mozambique but in the rest of the southern African region.

NGOs address every conceivable issue and they operate in virtually every part of the globe. NGOs are often seen as synonymous with non-profits, but they are different. Non-profits include a very wide range of organizations including museums, universities, and hospitals, that focus on services and rarely (if ever) engage in advocacy.[49] By contrast, NGOs always have an important advocacy mission. An example of one of the larger NGOs is Amnesty International. They are based in London, have more than a million members, and have affiliates or networks in over 90 countries and territories.

NGOs are very well respected around the world. For example, an opinion poll in Germany found that considerably more respondents said they trusted the NGO Greenpeace than those that expressed trust in the German Federal government.[50] Additionally, IGOs do often try to weed out the bad from the good. International organizations like the UN require officially accredited NGOs to pass through a review process to determine which are legitimate partners, and in recent years this collaboration of NGOs and IGOs has led to some important accomplishments. This included strengthening women's rights and reaching agreements that help the environment (see *International agreements* below). Some believe that these successes resulted from increasing globalization and the pressure of ordinary citizens to control and regulate the world beyond the nation state.[51]

As the world's population grows and as the planet continues to shrink, NGOs will no doubt play a very large role in the future. Discussions about democracy and accountability in global decision-making will continue and NGOs will be there for the people. James Paul, Executive Director of the Global Policy Forum states, "Globalization has created both cross-border issues that NGOs address and cross-border communities of interest that NGOs represent. National governments cannot do either task as effectively or as legitimately. In the globalizing world of the 21st century, NGOs will have a growing international calling."[52]

Indonesia, January 31, 2005—With the help of U.S. Air Force helicopters, the World Food Programme (WFP) delivers the first canned fish and vegetable oil to the coastal town of Lamno on the west coast of Aceh. Staff members of WFP and the NGO partner *Action Contre La Faim* (Action Against Hunger) were waiting with a team of loaders and a ten-ton truck.

## International agreements

In the international politics arena there have been some remarkable achievements in the past few decades. Some of these feats have been highlighted in previous chapters like the quest of Jody Williams to ban landmines (Chapter 6). At the end of these great stories usually lies an international agreement; however it must be understood that there are differences in the terms dealing with international agreements.

Presidents Putin and Bush sign the Moscow Treaty to reduce nuclear stockpiles, May 2002.

There are many different names for agreements entered into by actors under international law. The names include: treaties, international agreements, protocols, conventions, exchanges of letters, exchanges of notes and so on. No matter what the name is, all of these international agreements under international law share the same rules. Wikipedia loosely compares them to contracts: both are means of willing parties assuming obligations among themselves, and a party to either that fails to live up to their obligations can be held liable under international law for that breach.[53]

There is also a difference between bilateral and multilateral treaties. Bilateral treaties are usually negotiated between two states. They establish the obligations and legal rights between only those two states involved. Multilateral treaties, which I highlight in this section, are usually open to any state and involve many parties. Multilateral treaties establish rights and obligations between each party and every other party.[54]

In international law, you often read about protocols such as the Kyoto Protocol. Wikipedia states, "A protocol is a treaty or international agreement that supplements a previous treaty or international agreement. A protocol can amend the previous treaty, or add further provisions. Parties to the earlier agreement are not required to adopt the protocol; sometimes this is made clearer by calling it an 'optional protocol', especially where many parties to the first agreement do not support the protocol." Using Kyoto as an example, it is a protocol to the United Nations Framework Convention on Climate Change (UNFCCC) which deals with greenhouse gas emission limits, while the Kyoto Protocol contained the specific provisions and regulations later agreed upon.[55]

The UN also plays a role in international agreements. The UN Charter states that a treaty must be registered with the UN before it can be enforced in its judiciary organ, the International Court of Justice.[56] This was originally put into the Charter owing the secret treaties that took place in the 19th and 20th centuries.

## Major international agreements (all subjects)[57]

| Year | Agreement |
|---|---|
| 1925 | Protocol for the Prohibition of the Use of Asphyxiating, Poisonous or Other Gases, and Warfare (Geneva Convention), Includes Convention I, II, III, and IV. Convention (I) for the Amelioration of the Condition of the Wounded and Sick in Armed Forces in the Field. Convention (II) for the Amelioration of the Condition of Wounded, Sick and Shipwrecked Members of Armed Forces at Sea. Convention (III) relative to the Treatment of Prisoners of War. Convention (IV) relative to the Protection of Civilian Persons in Time of War. |
| 1948 | Universal Declaration of Human Rights. |
| 1950 | Convention for the Suppression of the Traffic in Persons and of the Exploitation of the Prostitution of Others. |
| 1951 | Convention Relating to the Status of Refugees. |
| 1954 | Convention for the Protection of Cultural Property in the Event of Armed Conflict. |
| 1959 | Antarctic Treaty (no military use and nuclear testing). |
| 1966 | International Covenant on the Elimination of All Forms of Racial Discrimination. |
| 1966 | International Covenant on Civil and Political Rights. |
| 1967 | Protocol Relating to the Status of Refugees. |
| 1967 | Outer Space Treaty (military weapons). |
| 1968 | Nuclear Non-proliferation Treaty (NPT). |
| 1972 | Convention on the Prohibition of the Development, Production and Stockpiling of Bacteriological (Biological) and Toxin Weapons and on their Destruction. |
| 1976 | Convention on the Prohibition of Military or any other Hostile Use of Environmental Modification Techniques. |
| 1979 | Convention on the Elimination of All Forms of Discrimination against Women. |
| 1980 | Convention on Prohibitions or Restrictions on the Use of Certain Conventional Weapons which may be deemed to be Excessively Injurious or to have Indiscriminate Effects. |
| 1984 | Convention against Torture and Other Cruel, Inhuman or Degrading Treatment or Punishment. |
| 1987 | Montreal Protocol on Substances Depleting the Ozone Layer. |
| 1990 | International Convention on the Protection of the Rights of All Migrant Workers and Members of their Families. |
| 1991 | United Nations Framework Convention on Climate Change. |
| 1992 | Convention on the Prohibition of the Development, Production, Stockpiling and Use of Chemical Weapons and on their Destruction. |

| | |
|---|---|
| 1995 | Agreement for the Implementation of the Provisions of the United Nations Convention on the Law of the Sea of December 10, 1982 relating to the Conservation and Management of Straddling Fish Stocks and Highly Migratory Fish Stocks. |
| 1995 | Protocol on Blinding Laser Weapons (Protocol IV to the 1980 Convention). 1996 Protocol on Prohibitions or Restrictions on the Use of Mines, Booby-Traps and Other Devices (Protocol II to the 1980 Convention). |
| 1996 | Comprehensive Test Ban Treaty (CTBT). |
| 1997 | Convention on the Prohibition of the Use, Stockpiling, Production and Transfer of Anti-Personnel Mines and on their Destruction. |
| 1997 | Protocol to the United Nations Framework Convention on Climate Change (Kyoto). |
| 1998 | Rome Statute of the International Criminal Court. |
| 2000 | United Nations Convention against Transnational Organized Crime. |
| 2001 | Protocol against the Illicit Manufacturing of and Trafficking in Firearms, their Parts and Components and Ammunition, Supplementing the UN Convention against Transnational Organized Crime. |
| 2003 | United Nations Convention against Corruption. |

## Multinational corporations

AIDS activists target Coca-Cola at New York Gay Pride Parade demanding Coke pay for AIDS drugs for workers in poor countries.

Multinational Corporations (MNCs) can be defined as business enterprises headquartered in one state that invest and operate extensively in other states.[58] These corporations are all around us with Nike and Coca-Cola being two famous examples. Since globalization took off after World War II, MNCs have grown in size and importance. You will find that many MNCs have more money than well-established countries. With this economic globalization has come a bad stereotype for most of the MNCs, but is it all true?

One common perception is that the corporations pay very unfair wages. On the contrary, professor from Columbia University, Jagdish Bhagwati found just the opposite. Studies have been done in Bangladesh, Mexico, Shanghai, Indonesia, Vietnam, and elsewhere. They found that MNCs actually pay a 'wage premium', that is, an average wage that exceeds the going rate in the area where they are located. The studies also showed that, "affiliates of some U.S. multinationals pay a premium over local wages that ranges from 40 to 100 percent."[59]

A specific study on Vietnam was done by an economist named Paul Glewwe, using Vietnamese household data for 1997-98. He was able to isolate the incomes of workers employed in joint ventures, foreign-owned firms, and Vietnamese-owned enterprises. "About half the Vietnamese workers in the study worked in the foreign textile or leather firms that are so often criticized."[60] Glewwe found that workers in foreign-owned enterprises generally make almost twice the salary of the average worker employed at a Vietnamese company. Furthermore, Glewwe's data shows that "people who obtained employment in foreign-owned enterprises and joint ventures in Vietnam in the 1990s experienced increases in household income (as measured by per-capita consumption expenditures) that exceeded the average increases for all Vietnamese households."

There is also the account of two *New York Times* journalists Nicholas Kristof and Sheryl Wudunn who went to Asia to live. They were outraged when they first arrived at the sweatshop conditions. Asian factory workers were working sometimes 14 hours a day, six or seven days a week. But what they soon found was that factories were an important part of a "developing nation's journey to prosperity." One interview further elaborates: "One of the half-dozen men and women sitting on a bench eating was a sinewy, bare-chested laborer in his late thirties named Mongkol Latlakorn. It was a hot, lazy day, and so we started chatting idly about the food and, eventually, our families. Mongkol mentioned that his daughter, Darin, was 15, and his voice softened as he spoke of her. She was beautiful and smart, and her father's hopes rested on her.

"Is she in school?" we asked.

"Oh, no," Mongkol said, his eyes sparkling with amusement. "She's working in a factory in Bangkok. She's making clothing for export to America." He explained that she was paid $2 a day for a nine-hour shift, six days a week.

"It's dangerous work," Mongkol added. "Twice the needles went right through her hands. But the managers bandaged up her hands, and both times she got better again and went back to work."

"How terrible," we murmured sympathetically. Mongkol looked up, puzzled. "It's good pay," he said. "I hope she can keep that job. There's all this talk about factories closing now, and she said there are rumors that her factory might close. I hope that doesn't happen. I don't know what she would do then."[61]

This mobility is one of the main negatives against MNCs. Today many MNCs choose to move production abroad rather than deal with many of issues back home. Production is normally moved to developing countries that want the revenue and taxes and are not particularly interested in the high-tech legislations against environmental damage.[62] In addition, new companies receive tax breaks for moving into other countries. So because it is so simple just to move locations, MNCs have become a disadvantage to well-meaning governments. Food historian Colin Spencer discusses an old example in food industry and concludes that only a European Commission (EC)-wide regulation can place control of industry back into the hands of legitimate legislative authorities.

"[In] 1990 the U.K. banned the use of the veal crate; no longer could calves a few days old be imprisoned in a crate so small they could hardly move and be fed a liquid deficient in iron and fiber for those five months before slaughter. Instead the calves are exported out of the U.K. and imprisoned in the crates in other EC countries only to return as veal to the U.K.. [...]"[63]

Another big negative against MNCs is the lack of oversight. Author Naomi Klein writes: "Corporations have grown so big they have superseded government. That unlike governments, they are accountable only to their shareholders; [we] lack the mechanisms to make them answer to a broader public. There have been several exhaustive books chronicling the ascendancy of what has come to be called 'corporate rule'."[64]

One famous example is the case of Nike in Mexico. In 2001, a story broke regarding Nike and abuses at their factory in Kukdong, Mexico. Barra explains that larger American MNCs, such as Nike, often manufacture their product, in this case apparel, in foreign factories in order to save money on labor. This is no secret, but Barra further notes that, "these factories have often been in remote locations and away from the public eye, allowing for human rights abuses and violation of international law as dictated by the United Nations and other IGOs."[65]

Workers stitching uppers in a Siam Unisole factory. Manufacturers provide cloth-masks to workers to satisfy the codes of conduct. However, as much of the airborne pollutants in the factories are chemical and solvent-based, these masks are of little use.

On January 3, 2001, five workers were fired after trying to mobilize a protest because of unfair working conditions. This showed the failure of the Revolutionary Confederation of Workers and Peasants: their union which, "dedicated itself to the subordination of the worker to the factory owners and never pushed for more than minimum wage or minimum health standards."[66] A week later, 800 workers participated in work stoppages to protest this action and to call for a more legitimate, worker-focused union. Riot police violently broke up these strikes, and many of the workers were fired for their involvement. The factory workers had no support from other factory workers or union leaders so they asked for international help.

They contacted a non-profit organization called the Worker Rights Consortium (WRC). The WRC was created by college and university administrations, students and labor rights experts. Their purpose is to assist in the enforcement of manufacturing codes of conduct adopted by colleges and universities. These codes are designed to ensure that factories producing clothing and other goods bearing college and university names respect the basic rights of workers. The WRC investigated the plant at Kukdong and found several infractions of Mexican and international law including an illegitimate union, a violation of freedom to assemble, and numerous violations of human rights and health standards. The WRC also utilized its resources to raise public awareness about this small labor dispute in the U.S. and Mexico. It organized student groups on over 80 campuses which all got involved in protests, and by the end of January, articles were run in dozens of newspapers.

Once the word got out, other NGOs investigated, including the International Labor Rights Fund, an organization "dedicated to achieving just and humane treatment of workers worldwide." In the middle of March, Nike released a report that included a "detailed account of abuses that were alleged by factory workers, the veracity of these reports as determined by Nike's own investigating committee, and a timeline for fixing these issues."[67] Due to everyone's involvement, the factory is now under more favorable conditions. "They allowed workers to form their own union; they improved working condition and treatment of workers; they received two general pay raises; and working conditions now strictly adhere to international law."[68]

Even though there was a happy ending in this case, MNCs have real control over the host countries they operate in. The UN Committee on Trade and Development understands the importance of the host countries and MNCs.

Again, a key concern with regards to MNCs is their mobile nature. "In their negotiations with the government of the host country, their ability to pick up and leave provides them with a great deal of leverage over states dependent on the jobs they provide."[69] Host governments do have some bargaining power, but, particularly in developing nations, where economies are often weak, the concerns of the host government over how MNCs operate in their country must often take a backburner to investment concerns. That's where the risk of exploitation lies.

Dr. Gary M. Quinlivan of Saint Vincent College writes, "Many religious leaders are increasingly troubled by the growing presence of multinational corporations around the world, especially in poor and developing nations."[70] He believes concern is warranted. "Such allegations include the charge that profit-motivated multinational corporations are engaging in destructive competition and insidious plots to manipulate entire economies." Furthermore, MNCs are perceived to be using their size and power to eliminate domestic companies, exporting high-wage jobs to low-wage countries, destroying the world's environment, perpetuating world poverty, and exploiting child labor."[71]

MNCs can be a very positive influence if they are watched closely, but cases like Kukdong demonstrate that they often can avoid the laws of any one particular government. If a monitoring system is put in place, it would relieve the pressure on civil society and strong national legislation and enforcement as the sole options for regulating MNC behavior. IGOs like the UN and EU can fight back against MNCs and provide the oversight that is required in order to reign in MNCs that are presently beyond the law. David Held writes: "Nation-states, some argue, are too small to be able to influence global change, and too large to respond effectively to the pressures for increased flexibility and competitiveness, or as Giddens put it 'too small to solve the big problems, but also too large to solve the small ones'."[72]

## The remaining problems

Andromeda Tower and Vienna International Center (VIC) in Vienna, Austria. It is also known as UNO City in Vienna and is the campus and building complex hosting the following UN organizations: International Atomic Energy Agency (IAEA); United Nations Office on Drugs and Crime (UNODC) and UN Office at Vienna (UNOV) including Office for Outer Space Affairs (OOSA); United Nations Industrial Development Organization (UNIDO); and the Preparatory Commission for the Comprehensive Test Ban Treaty Organization (CTBTO).

As we come close to the end of the first decade of the 21st century, we question whether independent states will be able to mold a peaceful future.

Right now it appears they cannot. The problems discussed throughout Part I will continue until there is a true collective effort by all. It appears the world is headed in that direction, but the last bump in the road will definitely be the most difficult.

Right now we can see faults and obstacles in all the international organizations mentioned in this chapter. Starting with the UN, although their accomplishments in all aspects of global issues have been substantial, the future of the organization remains unclear. Addressing this uncertain future, Kofi Annan warned that, "The global agenda has never been so varied, so pressing or so complex. It demands of the international community new approaches, new resources and new commitments of political will."[73] In order to get to this 'new approach' we have to deal with the issue of sovereignty.

Sovereignty basically means that no authority is legally above the state expect that which the state voluntarily confers on the international organizations. Due to this, many states will have an extremely difficult time adhering to international laws. This is the major predicament as it highlights the tension regarding national rights versus global responsibilities. Kegley and Wittkopf state, "Globalization captures the idea that everyone and everything on the planet is now more closely connected than ever before, but on an institutional foundation that is shaky and unprepared for managing the massive adjustments brought on by accelerating worldwide changes. Most affected and at a loss are the so-called sovereign states, no longer able to exercise control over the economic, political and environmental problems within their territories. States everywhere are experiencing a sense of lost sovereignty as they attempt to cope with the pressures originating outside their borders."[74]

Looking at the ICC you can see this problem of lost sovereignty. The creation and existence of the ICC has been controversial with a number of states mainly due to the source and nature of the court's jurisdiction. Some countries object to the court, saying that there is very little legal supervision, while others think the court's verdicts may become subject to political motives.

The U.S.'s main objection to the ICC is national sovereignty and a fear of politically motivated prosecutions. They have not ratified the treaty creating the court, and do not intend to do so because, as former Congressman Tom Delay stated, "The ICC presents a clear and present danger to the war on terror and to Americans that are fighting it all over the world."[75]

In August of 2006, the island nation of Nauru and the Balkan nation of Montenegro signed the Geneva Conventions of 1949 meaning all the nations of the world have now signed the treaty.[76] This is supposed to ensure humane treatment in conflict, embodying centuries of law, custom, and the practise of placing warfare under the rule of law. David Crane states that, unfortunately, the U.S., once the champion of the Geneva Convention, committed breaches of their provisions in Iraq, Afghanistan, Guantanamo, and Eastern Europe in its single-minded conduct of the war on terror.

Mr. Crane points out: "Since the Geneva Conventions nations have entered into numerous agreements limiting or outlawing such things as torture, genocide, the use of bacteriological and chemical weapons to name a few. Key human rights instruments have also been established, such as the Universal Declaration of Human Rights. The creation of important institutions like the United Nations, the International Court of Justice, the International Criminal Court, the various international criminal tribunals, as well as the rise of non-governmental institutions and civil society allow for mechanisms to be created to seek justice and monitor impunity. Yet, despite all of this, the murder, rape, maiming, mutilation, pillaging, destruction of non-combatants worldwide goes on unabated. In such places as Darfur, Iraq, the DRC, Uganda, and Colombia human suffering continues."[77]

What can we do to stop these atrocities from occurring? It is clear that this issue of national sovereignty is an enormous stumbling block. As Hindell argues, "Taking a slightly longer-term view, the issues of climate change, environmental pollution, AIDS, migration, drugs, and international crime all require some kind of supranational authority to act within the boundaries of the [nation] state."[78] Some argue that without a supranational authority to enforce these international agreements, individual nations will allow suffering and pain to occur. "If national sovereignty resists the measures to reverse climate change, some UN members will drown while others could lose large slices of their territory."[79] Others wonder what the good is of an international criminal court if it is not backed up by an international law-enforcement agency with powers of arrest, detention, arraignment, trial and imprisonment.

Right now it appears we are headed in that direction of global governance but are not able or willing to take that final and most difficult step. With institutions being created like the ICC and the World Court, is that next and final step a world government? Can governments, civil society and the private economic sector come together to take collective action to address global issues? Is a supranational authority really the answer and what does the future hold?

# Part 3

# Is global governance the answer?

# Chapter 8

# Global governance—its history and future benefits

"Imagine all the people living life in peace. You may say I'm a dreamer, but I'm not the only one. I hope someday you'll join us, and the world will live as one."
—John Lennon, musician

Framed by the Earth's horizon and airglow, the full moon floats in the blackness of space in this photo from the Expedition 10 crew on board the International Space Station. The station plays a crucial role as an orbital outpost, as NASA pursues an exploration strategy that will send humans back to the lunar surface and eventually on to Mars and beyond.

Although many groups and individuals are working hard to cure the world's problems and to bring about world peace, does more have to be done to obtain this goal? Some people feel that the answer to all of our problems might be a single global government. This remedy is often referred to as 'world federalism', or a school of thought advocating the merger of previously sovereign states into a single integrated union.[1]

If this integration were to take place, it would not mean that states would be abolished and replaced with some central authority. The federalists reason that if people truly value the conditions in which all participants become better off more highly, then the conditions in which some benefit more than others (such as an individual state's national advantage over its rivals), they will readily transfer their loyalty to a supranational authority to dismantle the anarchical system of competitive territorial states that produces war.[2] All in all, federalists believe that people are good, wise and enlightened, and that a world government is the answer.

I feel that people do know peace is better than war. The people who see the bigger picture understand the fact that we share only one atmosphere and a handful of oceans. One of these people who shares this view is Elizabeth Sawin. She writes, "We know that to steward a global resource we need global agreements, and global bodies to implement them. We know we must speak with one voice about climate, about water, about refugees, and about poverty. We are one people on one world and most of us know this, somewhere inside of ourselves, even when that is not what we have been taught, or what the people who govern us proclaim."[3] Elizabeth Sawin also sees the beauty of the words that have come from various astronauts who have seen our planet from space. Their words are so beautiful that you will see them scattered throughout this book.

When Russell Schweickart orbited Earth in 1969 he said: "You look down there and you can't imagine how many borders and boundaries you cross, again and again and again, and you don't even see them. There you are—hundreds of people in the Middle East killing each other over some imaginary line that you're not even aware of, that you can't see. And from where you see it, the thing is a whole, the earth is a whole, and it's so beautiful."[4]

Sawin states that those words touch us not only because they were inspired by a journey to space, but also because we can feel their truth down here on the ground. These words of various astronauts show us the reality of the wholeness of our planet. It is something that is harder and harder for us to ignore and soon we are going to identify ourselves as citizens of the whole planet. Additionally, "More and more of us are going to be turning to something that we have yet to create—a government that can hear all of us, all of the people of a world that is whole."[5]

## New World Order conspiracy

"All truth passes through three stages. First it is ridiculed. Second, it is violently opposed. Third, it is accepted as being self-evident."—Schopenhauer, German philosopher

'Black Helicopters': part of the U.S. militia conspiracy theory. They claim that special unmarked 'black' helicopters are used by secret agents of the New World Order. In the U.K. a similar phenomenon known as 'Phantom Helicopters' has been reported since the mid-1970s. The helicopters pictured here were used as props in a science fiction movie.

It is important to note that there are many people who are very much against the idea of a world government. One of the main conspiracies against this idea is commonly known as the New World Order. According to Wikipedia it refers to a conspiracy theory in which a powerful and secretive group (Illuminati, Freemasons, etc.) has created a secret plan to rule the world eventually via a unitary (as opposed to a federal) world government. These theorists believe that historical and current events are seen as steps in an ongoing plot to rule the world. They cannot say who is part of this plot but they feel prominent families such as the Rockefellers are important members. Presidents and prime ministers of nations are routinely included in this huge conspiracy.

The conspirators thought to be responsible for the New World Order are also suspected of staging many historical events such as UFO sightings and large-scale wars. Some prominent examples include:

- The U.S. supposedly knew in advance of the attacks on Pearl Harbor and President Roosevelt used the attacks as a 'legitimate' reason for entering World War II.
- The Gulf of Tonkin incident which led President Johnson to escalate U.S. hostilities in Vietnam.
- The September 11, 2001 attacks, engineered by elements of the conspiracy, was a sacrifice the U.S. government deemed acceptable, since it allowed them to declare war on Afghanistan, and eventually Iraq, a large source of oil. This in turn would let the U.S. draw their oil under the façade of anti-terrorism measures, giving the conspirators even more power.[6]

## The Illuminati

The Illuminati, a secret society, was created in 1776. Since then, there is supposedly a master plan to bring about the subjection of the whole human race to an oligarchy of Financiers. In fact, the word 'Illuminati' is derived from Lucifer and means 'holders of the light'.[7]

Funnily enough, the insignia of the Order of the Illuminati appears on the reverse side of U.S. one-dollar bills. If you look closely at the base of the 13-storey pyramid you will see the year 1776 written in Roman numerals (MDCCLXVI). Most people think that it represents the date of the signing of the American Declaration of Independence, but it actually memorializes the foundation of the Order of the Illuminati, which was on May 1, 1776.[8]

According to one website: "The pyramid represents the conspiracy for the destruction of the Catholic Church (and all existing religions), and the establishment of a one-world government, or a United Nations dictatorship. The eye radiating in all directions, is the 'all-spying' eye that symbolizes the terroristic, Gestapo-like, espionage agency that Weishaupt set up to guard the 'secret of the Order and to terrorize the populace into the acceptance of its rule". The Latin words 'ANNUIT COEPTIS' mean "our enterprise (conspiracy) has been crowned with success". Below, 'NOVUS ORDO SECLORUM' explains the nature of the enterprise: a 'New Social Order" or a "New World Order'."[9]

In 1933, the appearance of this insignia was officially put on the U.S. one-dollar bill. Many feel that this was a sign that the Illuminati totally controlled the U.S. government. Indeed this quest for a world government has been going on for a long time. For example, on February 17, 1953, Financier Paul Warburg said before the U.S. Senate: "We shall have world government whether or not you like it, by conquest or consent."[10]

This conspiracy theory can at times seem very ridiculous to most of us, but there are those who believe that it is true. The fact of the matter remains: none of this theory is proven and the New World Order is a 'conspiracy', nothing more. However, there is truth in the statement that world government is by no means a new idea. In past decades and even centuries, various individuals have written novels, produced television shows and directed movies all using the notion of global governance. Let's look at some of the more popular world governments that have been used in the entertainment industry.

## World government in entertainment

The flag of the United Federation of Planets from the hit television series *Star Trek*.

To the average person, the most famous example of global governance would come from the hit television series *Star Trek*. In fact, *Star Trek* was and still is so popular that Wikipedia, the famous online encyclopaedia, devoted a whole section to the information contained in the show.

According to Memory Alpha, it was the year 2063 when first contact was made with extraterrestrials. The world government was formed some time in the 22nd century.[11] The government was given the name of 'United Earth' and continued throughout the whole show. According to the official *Star Trek* website, 2113 might have been the year that United Earth came into being. We know from episodes of *Star Trek: Enterprise,* mainly 'first flight' that the United Earth Starfleet existed in the 2130s and 2140s.[12]

What led up to the world government was that of a European hegemony which existed into the 2120s. According to *Star Trek: The Next Generation*, the last of Earth's states joined in 2150. However many states retained their individual identities. For example, the European Alliance, African Confederation and the U.S. lasted well into the 24th century and beyond. It is unknown how the world government was structurally set up. On two episodes of *Star Trek: Enterprise*, an Earth diplomat named Nathan Samuels appeared and was called by the name 'minister', which suggests that Earth may have had a parliamentary form of government. Mr. Samuels might have been the prime minister, or the leader of the Earth, but this is unproven. The United Earth Government, which ran the Earth Starfleet, remained in place following the founding of the United Federation of Planets.[13]

### *Starship Troopers*

Many young individuals have the seen the recent Hollywood blockbuster remake of the novel *Starship Troopers*. Starring Denise Richards, the movie's background is a military world government. This came about after the 'civilian' governments of the world proved their weakness for starting wars over silly things (like tariffs, human rights violations, oil, being of a different race, religion), and the military, tired of dying for greedy civilians staged a coup and took over the world, thus ensuring 'peace'.[14]

This military government of *Starship Troopers* is very different from what you are probably thinking. It is not a police state. This is extremely important to the story's theme. One can still strive to make a lot of money, study whatever one pleases at university, or worship as one so desires. Private individual rights are still very much intact. There is no racism or sexism and the leader of the world government (known as the 'Sky Marshall') is a woman of color.[15] Crime is very low and there is an all-volunteer military service. Those who serve for two years earn the right to participate in government and are awarded citizenship. Then one day this one Earth government gets attacked by a ruthless intelligence of basically giant bugs from a neighboring galaxy. All in all, the war in *Starship Troopers* is not of conquest, but of survival depending on thwarting an enemy dedicated to eradicating the human species. The true story of *Starship Troopers* is about a group of high school seniors coming of age under a military government.[16] It is a time in the future where children have ambitions of greatness in science and engineering and other professions that will benefit humanity, rather than rap music or becoming the next Michael Jordan. It is a pleasant thought to think about a government that rewards the best students with free college and the chance to participate in space exploration. This is definitely better than graduating and only receiving the horrible reward of paying back student loans.

### Other world government themes in entertainment

*The Truth Machine*: This 1996 book takes place in the future after the invention of a device which is much more reliable than conventional lie detectors. It becomes extremely popular all over the world and is available to everyone so that lying has become something which just does not pay any longer and, consequently, is generally considered erratic behavior which used to occur in the past.[17] Due to this machine, crime has become almost non-existent. Also, it makes a huge impact on politics. Those who really want to do well and truly want the best replace the corrupt politicians. As the only remaining superpower, the U.S. takes the lead and all the other nations readily join in to form a truly global society.[18]

*Left Behind*: This series of novels by Tim Lahaye tells us the story of the Rapture according to evangelical Christian theology. These books took the U.S. by storm and have sold over 50 million copies. According to Wikipedia, the stories center around the quick rise to power of a charismatic, populist Romanian politician named Nicholae who is eventually appointed Secretary-General of the UN. Once in power, Nicholae advocates a series of popular proposals, such as a global currency and massive international disarmament. Eventually, all countries unite into one 'Global Community'. Nonetheless, Nicholae introduces more and more totalitarian measures to unite the world, such as allowing only one legal religion, and only one global media outlet.[19] He is eventually seen as the antichrist and is rejected by the books' Christian protagonists. The idea that the antichrist will be the leader and chief advocate of a world government is a popular theme among believers in Christian Rapture theory.[20]

In a BBC interview, Tim Lahaye stated, "Never in the history of mankind has there been so much fear of self-annihilation ... People are asking the question what is going to happen in the future, and the Bible gives them the answer." Mr. Lahaye believes that as the end of the world approaches, it will be presided over by a world government. "The Bible is very clear that in the last days there is a world government—it could be the United Nations, because the UN is totally an anti-Christian organization, and has been from the beginning," he added.[21]

## World government in education

As early as the 1700s, many thinkers believed that all human beings are born equal and should enjoy certain basic liberties and rights including free elections. Some of the more famous names include Jean Jacques Rousseau from France, Immanuel Kant from Germany, Thomas Paine and Thomas Jefferson from the U.S., and David Hume from Scotland. Of these, Kant's main theory is still the best-known. In 1795, Kant advocated a "peaceful league of democracies." In his essay, 'Perpetual

Immanuel Kant from Germany.

Peace: A Philosophical Sketch,' Kant postulates three basic requirements for organizing human affairs to abolish permanently the threat of a future war:

- The civil constitution of each state shall be republican;
- The law of nations shall be founded on a federation of free states;
- The rights of people, as citizens of the world, shall be limited to the conditions of universal hospitality (i.e., people would be allowed to visit other countries, but not to stay unless invited).[22]

In 1842, the English poet Lord Alfred Tennyson, published the often-quoted lines ('Locksley Hall'): "For I dipt into the future, far as human eye could see / Saw a Vision of the world, and all the wonder that would be /... / Till the war-drum throbb'd no longer / and the battle-flags were furled / In the Parliament of man, the Federation of the world. / There the common sense of most shall hold / a fretful realm in awe / And the kindly earth shall slumber / lapt in universal law." Even American Ulysses S. Grant commented, "I believe at some future day, the nations of the earth will agree on some sort of congress which will take cognizance of international questions of difficulty and whose decisions will be as binding as the decisions of the Supreme Court are upon us."[23]

Unfortunately it has taken the events in our own wondrous and terrible 20th and 21st centuries to clinch the case for world government. With the arrival of electricity, air travel, television and the Internet, the planet has become smaller than ever, and it continues to shrink every passing day. In the beginning of the 20th century we learned that the price of settling international disputes by force was way too high, with the Hiroshima and Nagasaki being the breaking point. Once again great minds shared similar thoughts. Individuals such as Albert Einstein, Ghandi and Albert Toynbee all wrote in favor of interests higher than those of the nation.

Einstein wrote: "A human being is a part of the whole called by us universe, a part limited in time and space. He experiences himself, his thoughts and feeling as something separated from the rest, a kind of optical delusion of his consciousness. This delusion is a kind of prison for us, restricting us to our personal desires and to affection for a few persons nearest to us. Our task must be to free ourselves from this prison by widening our circle of compassion to embrace all living creatures and the whole of nature in its beauty."

Arnold Toynbee was more specific. His famous cyclical theory of history from 1954 explains: "The most emphatic cycle after another is the outbreak of a great war in which one power that has forged ahead of all of its rivals makes so formidable a bid for world domination that it evokes an opposing coalition of all the other powers."[24] Toynbee has indeed been proven right: the League of Nations in the 1920s, the UN in the 1940s and Western Europe creating NATO out of fear of the USSR in 1949.

Today, there are hundreds of international organizations which make us very much a united world. Most the globe's nearly 30,000 NGOs are sympathetic to peace plans that seek to reduce the capability of states to make wars.[25] Some organizations such as the United World Federalists and the World Future Council (see Chapter 10) are promoting the creation of a world government.

## Why global governance in the first place?

"Unless we establish some form of world government, it will not be possible for us to avert a World War III in the future ... Unless some effective world super-government for the purpose of preventing war can be set up ... the prospects for peace and human progress are dark ... If ... it is found possible to build a world organization of irresistible force and inviolable authority for the purpose of securing peace, there are no limits to the blessings which all men enjoy and share."—Winston Churchill, former British Prime Minister

Conspiracies and world domination aside, many benefits would arise from having a world government. Instead of examining every single benefit, I would like to focus on the more prominent advantages. On the World Beyond Borders website are listed three main benefits of forming a world government. They are: 1) eliminating war; 2) the global environment, and 3) space exploration and human unity.

### Eliminating war

"I know not with what weapons World War III will be fought, but World War IV will be fought with sticks and stones."—Albert Einstein, scientist

Hiroshima, after the first atomic bomb explosion. This view was taken from the Red Cross hospital building about one mile from the detonation.

Proponents of global governance argue that forming a world government would restore order around the globe. The world government could make sure that nuclear weapons were truly banned and that terrorism would be diminished. The world would also be able to control other forms of warfare many leaders use to manipulate their people such as food, raw materials and finance. In an article by the Universal House of Justice it states: "The achievement of such ends requires several stages in the adjustment of national political attitudes, which now verge on anarchy in the absence of clearly defined laws or universally accepted and enforceable principles regulating the relationships between nations. The League of Nations, the United Nations, and the many organizations and agreements produced by them have unquestionably been helpful in attenuating some of the negative effects of international conflicts, but they have shown themselves incapable of preventing war."[26]

If a world government was allowed to enforce international laws such as the Universal Declaration of Human Rights and the Convention on the Prevention and Punishment of the Crime of Genocide, war would no longer dominate international relations. Then again some feel that having such a global powerful force in place would make it easy for a few to dominate the planet. This would certainly not be the case. All the states would have their right to maintain a certain level of armaments, especially for purposes of maintaining internal order within their respective dominions. They would need these weapons to enforce the international laws, but would also have the assistance of the federal world government if need be.

Today there is an army of men and women, drawn from virtually every culture, race and nation on earth, who serve the various agencies of the UN including peacekeeping operations. They represent, "a planetary 'civil service' whose impressive accomplishments are indicative of the degree of cooperation that can be attained even under discouraging conditions."[27] By forming a world government to eliminate war, permanent peace among nations could very realistically take hold. Without fearing the suspicious neighbor or even nuclear holocaust, political peace would flourish. Moreover, the outrageous amounts of money spent on weapons and other military items (*see* Chapter 5) could be diverted into other more important avenues such as education.

**Global environment**

For millions of years, Antarctica, the frozen continent at the southern end of the planet, has been encased in a gigantic sheet of ice. Recently, the Gravity Recovery and Climate Experiment (GRACE) satellite has been taking sensitive measurements of the gravity for the entire Earth, including Antarctica. Recent analysis of GRACE data indicates that the Antarctic ice sheet might have lost enough mass to cause the worlds' oceans to rise about .05 inches, on the average, from between 2002 and 2005.

Although this may not seem like much, the equivalent amount of water is about 40 trillion gallons, equivalent to the amount of water used by U.S. residents in three months. Uncertainties in the measurement make the mass loss uncertain by about 21 trillion gallons. The iceberg pictured below is a small part of the Antarctic ice sheet. The picture was taken on the Riiser-Larsen ice shelf in December 1995. Future research will likely focus on trying to understand the data better, take more data, predict future trends, and understand possible effects of these trends on the future climate of our entire home planet.

In the early 1990s then U.S. senator Al Gore wrote a book entitled, *Earth in the Balance: Ecology and the Human Spirit* and he called for the reorganization of society: "I have come to believe that we must take bold and unequivocal action: we must make the rescue of the environment the central organizing principle for civilization ... Adopting a central organizing principle—one agreed to voluntarily—means embarking on an all-out effort to use every policy and program, every law and institution, every treaty and alliance, every tactic and strategy, every plan and course of action—to use. In short every means to halt the destruction of the environment and to preserve and nurture our ecological system."[28]

It has been over 15 years since then, and Mr. Gore's warning is even more correct. There are still many problems with the world's environment and the the global environmental governance structure is inadequate to deal with the challenges the world faces today. Global governance would change this and it needs to change fast.

One of the great accomplishments of the 20th century space program has been the ability to view the environmental problems from above. From space, astronauts and satellites easily see the thinning ozone over the Antarctic. We know it is being eaten away by refrigerants and aerosol propellants used 20 years ago in London or Los Angeles. Co-founder of the World Beyond Borders website, Jane Shevtsov points out, "We see enormous dust clouds leaving China and know that agriculture that depletes the soil also pollutes the air hundreds or thousands of miles away. We see rainforest-burning in Brazil and know it makes the Arctic icepacks melt. Yet we persist in acting as if political boundaries reflected physical or biological reality."[29]

The current international organizations do not have the mission or resources to respond to global environmental externalities and to establish norms for behavior in international environmental relations. Their mandates are narrow, budgets small, and have limited support. There also exists a wide discrepancy between the funds spent to address environmental problems at the local, state/provincial, and national levels, on the one hand, and at the global level, on the other. Countries spend as little as 1/100th or even 1/1,000th of their environmental budgets on global and transboundary issues.[30]

This solution of global environmental governance is not new. Almost ten years ago, the Yale Center for Environmental Law and Policy Research Program met in New York to discuss strengthening global environmental governance. In the concluding paper from that particular meeting they laid out what the functions of such a global organization would be.

It states that in order to respond to pressing global-scale problems, a global environmental organization should have capacity to:

- gather data and monitor environmental indicators (publish periodic reports on the state of the global environment); provide an early warning system to forecast and identify environmental disasters and societies at risk;
- promote information and technology exchange (in the richer form often called 'knowledge networking')—taking advantage of the opportunity to achieve economies of scale in the technical, scientific, and analytic functions necessary for good environmental policymaking;
- serve as a catalytic force to focus the world community on emerging global-scale pollution and resource management issues;
- act as a convening authority to coordinate global responses to environmental challenges (engaging not only governments but also civil society at large—including business and non-governmental organizations interested in environmental issues);
- offer a forum for dispute resolution, and
- establish policy guidelines to promote global responses to environmental challenges.[31]

We must not forget that the environment is a global concern and it needs a global solution. We must not forget that two-thirds of the Earth is beyond the territorial sovereignty of UN member states; it is truly the 'global commons'.[32] Jane Shevtsov states, "Unenforceable standards are useless. In other words, we need global environmental laws made by a global government."[33] Indeed, global governance could very well be the answer to the problem we accidentally created many years ago.

**Space exploration and human unity**

Throughout our past, contemplating space has been one of the richest sources of insight into the fact that we are one species on a single planet.[34] In the late 1600s, the Dutch astronomer Christiaan Huygens published a book

Like a Roman candle shooting through the sky, the launch of Space Shuttle *Discovery* on mission STS-121 kicks off the fireworks for the U.S. independence day holiday. Commanded by Steven Lindsey, *Discovery* and its crew of seven astronauts roared from Launch Pad 39B to begin a 15,000-mph chase to rendezvous with the International Space Station.

that talked about other planets and the possibly of inhabitants. Huygens wrote, "How vast those Orbs must be, and how inconsiderable this Earth, the Theatre upon which all our mighty Designs, all our Navigations, and all our Wars are transacted, is when compared to them. A very fit consideration, and matter of Reflection, for those Kings and Princes who sacrifice the Lives of so many People, only to flatter their Ambition in being Masters of some pitiful corner of this small Spot."[35]

Three hundred years later, astronomer Carl Sagan echoed these same sentiments when he wrote of a picture of Earth from beyond Pluto's orbit: "Look again at that dot. That's here. That's home. That's us. On it everyone you love, everyone you know, everyone you ever heard of, every human being who ever was, lived out their lives. The aggregate of our joy and suffering, thousands of confident religions, ideologies, and economic doctrines, every hunter and forager, every hero and coward, every creator and destroyer of civilization, every king and peasant, every young couple in love, every mother and father, hopeful child, inventor and explorer, every teacher of morals, every corrupt politician, every 'superstar', every 'supreme leader', every saint and sinner in the history of our species lived there—on a mote of dust suspended in a sunbeam. The Earth is a very small stage in a vast cosmic arena. Think of the rivers of blood spilled by all those generals and emperors, so that, in glory and triumph, they could become the momentary masters of a fraction of a dot. Think of the endless cruelties visited by the inhabitants of one corner of this pixel on the scarcely distinguishable inhabitants of some other corner, how frequent their misunderstandings, how eager they are to kill one another, how fervent their hatreds."[36]

Space exploration, like astronomy in earlier centuries, provides a perspective from which to view human global affairs. This view makes a hundred or so warring nation-states seem absurd. But overall, this perspective, which is reinforced by photographs of Earth from space and the quotes of astronomers, brings with it a powerful argument for the political unification of our world. If you give space exploration any thought at all, you have to change your personal perspective from Earth-bound to that of no boundaries. With this change of perspective, "comes a feeling that space programmes ought to be undertaken by a united humanity."[37]

Looking back, most people who were alive will never forget July 20, 1969, when U.S. astronaut Neil Armstrong became the first human being to set foot on the moon. It took our species only 66 years to go from the birth of powered flight to landing on the moon. If you think about it, that is truly a remarkable feat. However, it is now been 38 years since that historic day and what do we have to show for it? It is unfortunate that the space station has never fired up the imagination of the public like President Kennedy's goal of putting man on the moon. Today the situation has become so dire that many people even question the need for NASA. The money allocated to the organization proves that point. NASA's budget has not increased significantly in many years. The 2007 budget request is about the same as

2006, just under US$15.5 billion. To put this in perspective, consider that the IRS has a budget of US$10.4 billion just to collect our taxes in the first place. Additionally, the Office of Homeland Security's budget for border control is only US$16.5 billion.[38]

Due to the questioning of NASA, many now believe that only a world government will be able to satisfy the political and economic conditions which must be met before humanity will be in a position to undertake any really large-scale space activities.[39] Mr. Ian Crawford has spoken about this topic on many occasions. His overall belief is that in large measure, focus is needed on the provision of resources for the development of space infrastructure, and the probability that only a disarmed world will be able to channel sufficient financial, material and intellectual resources in this direction. He adds, "As it seems certain that the world will not disarm until such time as a competent world government can guarantee international peace and security, such an institution would appear to be a prerequisite for large-scale space development."[40] To sum up, he and I would both argue that for any really large-scale program aimed at the exploration, utilization, and colonization of the solar system, a world government is needed.

However, NASA has started its quest to take man back to the moon and beyond. Although it will not even come close to the same scale as a true world space program, it is good step in the right direction. In 2006, NASA modified its mission statement by removing the phrase "to understand and protect the home planet" in an effort to re-focus the agency on outward-facing space exploration.[41] President Bush has already outlined his vision for NASA: go back to the moon and then to Mars. The next huge step would be to land a human being on the Martian surface. When there, we would be able to seek for life and answer one of life's big questions. As political author Tad Daley points out, "If the biochemistry made clear that Martian life derived from a separate and independent origin, it would strongly suggest that the universe is teeming with the stuff, and strongly imply that somewhere else it has evolved into sentience—that We Are Not Alone ... that would surely be the single greatest discovery in all of human history."[42]

Daley feels that isn't the main reason why we would venture to Mars, "But the real reason we should go is that Mars is the next step to ... 'Breaking out into the solar system.' Mars is the next stepping-stone on what the Planetary Society calls 'humankind's greatest adventure', the next stop on the infinite journey, the next rung on the ladder to heaven."[43] But why we would we want to construct bases on planets and moon and continue the journey deeper into space?

The only way to guarantee immortality for the whole human race is further space exploration. In a speech given on June 13, 2006, world-renowned scientist Stephen Hawking stated, "The survival of the human race depends on its ability to find new homes elsewhere in the universe because there's an increasing risk that a disaster will destroy the Earth."[44] He continued, "It is important for the human race to spread out into space for the survival

After driving a short distance from their landing site, two explorers stop to inspect a robotic lander and its small rover in this artist's concept of a future Mars mission. This stop also allows the crew to check out the life support systems of their rover and space suits within walking distance of the base.

of the species ... Life on Earth is at the ever-increasing risk of being wiped out by a disaster, such as sudden global warming, nuclear war, a genetically engineered virus or other dangers we have not yet thought of." He added that if humans can avoid killing themselves in the next 100 years, they should have space settlements that can continue without support from Earth.[45]

Nowadays it is easy to envision the world being wiped out by a cosmic event. Most of us have seen the movies *Armageddon* and *Deep Impact* which displayed such a scenario. Moreover, most living things have in fact been wiped out several times before by just such events in the past, most notably the supposedv dinosaur-killer asteroid over 65 million years ago. Daley adds, "God knows these days it's easier still to envision scenarios by which we foolishly annihilate ourselves. But if a thousand years from now we have established a durable presence in several locations far distant from one another—throughout the solar system and beyond—it's hard to imagine any apocalyptic event that could manage to eliminate us all. If *homo sapiens* can hang on until then, we will be as close to immortality as the universe itself."[46]

Since 9/11 it sometimes seems that the war on terror is everything. That the only purpose we have now is to delay another horrific attack. But the purpose of a civilization is not just to prevent destruction, but to engage in creation. If we let ourselves be deprived of that, then the terrorists surely do win. Coming together as one and putting all of the world's resources into a space program led by a world government would surely lead to advances and discoveries far beyond our time. Daley adds, "Perhaps we can demonstrate our true triumph over Osama bin Laden and his foul cohorts by going to Mars—and bringing a piece of the World Trade Center along."[47]

As stated earlier, astronauts of all nations have provided some of the most eloquent testimonials of our planet's wholeness and our species' unity. Michael Collins, who flew on Gemini and Apollo, said, "I really believe that if the political leaders of the world could see their planet from a distance of ... 100,000 miles, their outlook would be fundamentally changed. That all-important border would be invisible, that noisy argument suddenly silenced. The tiny globe would continue to turn, serenely ignoring its subdivisions, presenting a unified face that would cry out for unified treatment."[48]

Reflecting ecological concerns common to many astronauts, Yuri

Artyukhin said, "It isn't important in which sea or lake you observe a slick of pollution, or in the forests of which country a fire breaks out, or on which continent a hurricane arises. You are standing guard over the whole of our Earth."[49]

Apollo astronaut Edgar Mitchell was probably the most direct. "[In outer space] you develop an instant global consciousness, a people orientation, an intense dissatisfaction with the state of the world, and a compulsion to do something about it. From out there on the moon, international politics look so petty. You want to grab a politician by the scruff of the neck and drag him a quarter of a million miles out and say, 'Look at that, you son of a bitch'."[50]

# Chapter 9

# Global governance—how do we get there?

"A federation of all humanity ... would mean such a release and increase of human energy as to open up a new phase in human history."—H.G. Wells, writer

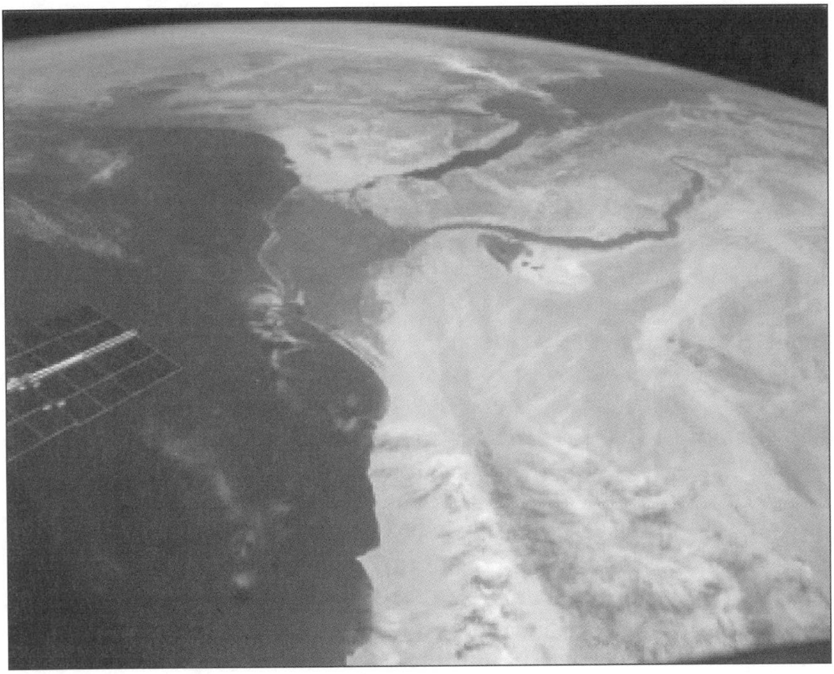

Egypt, Africa, and Mediterranean Sea.

Most believe that the best mechanism for democracy for the planet as a whole is not an all-powerful centralized super state, but a federation; a union of separate states that allocate certain powers to a central government while retaining many others for themselves.[1] So if this form of global governance is the answer, how do we get there? Is using the UN the answer? Should we look towards the EU for ideas? Is the U.S. a perfect example of how we should venture forward?

## The United States of America example

"The abolition of war is no longer an ethical question to be pondered solely by learned philosophers and ecclesiastics, but a hardcore one for the decision of the masses whose survival is the issue. Many will tell you with mockery and ridicule that the abolition of war can only be a dream—that it is the vague imagining of a visionary. But we must go on or we will go under. We must have new thoughts, new ideas, new concepts. We must break out of the straightjacket of the past. We must have sufficient imagination and courage to translate the universal wish for peace which is rapidly becoming a necessity into actuality."—General Douglas MacArthur, U.S. Army

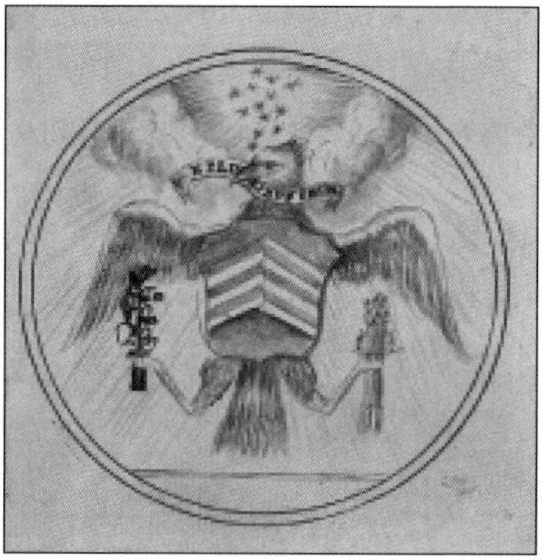
Original design of the Great Seal of the United States (1782).

For the most part, federalism is largely an American invention. In view of this, some feel that the U.S. is still the best example of a multinational federal state. Can we learn by looking back in time at America's Founding Fathers and their quest for a federal government?

When we think back to the late 1700s, most of us picture a smooth transition for America. We believe our Founding Fathers signed the Declaration of Independence, then the Constitution and kept expanding west. It sounds like a simple time but on the contrary, it was one of the most difficult times in U.S. history. Many years ago, Clarence K. Streit, former editor and League of Nations correspondent for the *New York Times*, wrote an article entitled, 'The American Example'. In his article he reminds us that at the beginning of 1776, independence was so remote from American thought that it was not even proposed seriously until January 10, when Thomas Paine came out for it. Thomas Paine himself showed us how difficult and remote the union of the 13 states really was when in 1787 he said, "If there is a country in the world where concord, according to common calculation, would be least expected, it is America. Made up as it is of people from different nations, accustomed to different forms and habits of Government, speaking different languages, and more different in their modes of worship, it would appear that the union of such a people was impracticable."[2]

A year earlier, in 1786 the country was a mess, to put it lightly. Trade disputes threatened war among New York, Connecticut and New Jersey.

Territorial disputes led to bloodshed and threat of war among New York, New Hampshire and Vermont, and between Connecticut and Pennsylvania. War with Spain threatened to break the League of Friendship in two camps. The League's Congress often had no money in the Treasury, and could no longer borrow. The States issued worthless currency, courts were broken up by armed mobs and misery was widespread. George Washington himself, America's first president, wrote in 1786, "I am uneasy and apprehensive, more so than during the war."[3]

At the time it seemed the "clashing interests of the Americans, their differences of governments, habitudes, and manners, indicate that they will have no centre of union and no common interest. They never can be united into one compact empire under any species of government whatever; a disunited people till the end of time, suspicious and distrustful of each other, they will be divided and sub-divided into little commonwealths or principalities, according to natural boundaries, by great bays of the sea, and by vast rivers, lakes, and ridges of mountains."[4] The situation was so bad that the 1787 Federal Convention had to be postponed because of a lack of representatives and interest needed from the 13 states. Even then the Convention was only to take place to reform the League and revise the Articles of Confederation. Only on the eve of the opening of the Convention did the idea of building a new document from scratch come to the table, thanks to George Washington's personal intervention.[5] Within 100 days and two years of close votes and heated debates, freedom-loving American democrats agreed to try out this invention on themselves. Twenty months after they read its text the American people established the Constitution that still governs them today. It is a remarkable accomplishment and an even better story, but what can we take away from it?

We as a planet are in a similar situation to that of the Americans in the beginning of their country. We come from very different backgrounds and share different beliefs. We must come together as the very first Americans did. We must acknowledge our global citizenship and loyalty to all of humanity. Streit wrote, "If that model does indeed work globally, it would be the logical extension of the Founding Fathers' wisdom, therefore a special source of pride for a world government's American constituents."[6] The Founding Fathers are considered 'men of vision', and we as a planet must follow in their footsteps as global governance is the next big step in our evolution.

## The European Union example

"It is quite clear that as long as the nations of the world spend most of their energy, money, and emotional strength in quarrelling with words and weapons, a true offensive against the common problems that threaten human survival is not very likely. A world government that can channel human efforts in the direction of the great solutions seems desirable, even essential. Naturally, such a world government should be a federal one, with regional and local autonomy safeguarded and with cultural diversity promoted."—Isaac Asimov, professor and author

The EU is a fine example of a supranational regional organization. Being far more recent than the U.S., their large and smooth transition can be seen as a terrific way for the world's nations to come together under a single administrative umbrella. The EU is the world's biggest free trade area with over a half a billion people and one single euro currency.[7] This collaboration is the first time in history that the continent has united as one.

Map of 'Euro-land' as it has become known. There are a total of 27 democratic member states in the EU. The most recent additions are Bulgaria and Romania who became part of the EU on January 1, 2007.

They not only have a common voice on economics, but also in defense and foreign policy. The EU has what is called its Common Foreign and Security Policy (CFSP). Its goal is to safeguard "the common values, fundamental interests, and independence of the Union."[8] It also strengthens each member's security, preserving "world peace and international security [as well as promoting] international cooperation to develop and consolidate democracy and the rule of law, and respect for human rights and fundamental freedoms."[9] In 2001, they also established the European Rapid Reaction Force. This enables them to have a presence throughout the world under one voice. This large military force also takes some of the strain off of the U.S. and NATO.[10]

Looking back at the history of Europe, the EU is an enormous achievement.

Since man first arrived on the continent thousands of years ago, Europe has been plagued with war and violence. Today there is a true sense of European citizenship, with the people living in Europe even having their own EU passports. The Europeans are also allowed to live, travel, work and invest in any of the EU member countries. It seems almost as if the EU could simply expand to engulf the rest of the world. But if only things were that simple.

In 1970, authors Roger Cobb and Charles Elder released their research that showed that the successful integration is a lot easier said than done. They found that it is a lot more than just two or more countries acting cooperatively. They need to share similar factors like geographical proximity, steady economic growth, similar political systems, supportive public opinion led by enthusiastic leaders, cultural homogeneity, internal political stability, similar experiences in historical and internal social development, compatible forms of government and economic systems, similar levels of military preparedness and economic resources, a shared perception of a common threat, bureaucratic compatibilities, and previous collaborated efforts.[11] Kegley and Wittkopf add, "While not all of these conditions must be present for integration to occur; the absence of more than a few considerably reduces the chances of success."[12]

To integrate the entire world into a system like the EU is not easily accomplished. And more importantly, it appears it will never carry that far. According to the EU website, the period 2007 to 2015 should see further enlargements of the EU. In the meantime, its leaders—listening carefully to public opinion—will have to decide where, ultimately, to draw the Union's geographical, political and cultural frontiers.[13]

## The United Nations example

"It seems to many of us that if we are to avoid the eventual catastrophic world conflict we must strengthen the United Nations as a first step toward a world government patterned after our own government with a legislature, executive and judiciary, and police to enforce its international laws and keep the peace."—Walter Cronkite, American broadcast journalist

Perhaps the best option in forming a world government is to utilize the international mechanisms already in place although we must be clear that the UN designed for the world of the 1940s is far from optimal for the challenges of the 21st century. To understand how to implement that change correctly, we must briefly look at the past to understand where the UN is today.

The original League of Nations collapsed for several reasons but mainly it was due to non-compliance from states and its weak structure.[14] Nations continued to make agreements based on their national interest. Due to this failure of the League of Nations to become a supranational authority, the UN was less idealistic in its goals, eventually focusing not on becoming a supranational authority but on developing mechanisms of regulating the

Rwandan peacekeepers return home from Darfur in October 2005.

balance of power between the two world blocks.[15] Now with the end of the Cold War, the UN is once again in a transition phase, with UN reform being very big news.

In 1995, the UN Commission on Global Governance wrote its report redesigning the UN and rewriting the UN Charter. Their final draft entitled 'Our Global Neighborhood', is a 410-page document which calls for dramatically strengthening the UN, by implementing a large list of recommendations, including these:

- Eliminating the veto and permanent member status in the Security Council;
- Authorizing global taxation on currency exchange and use of the 'global commons';
- Creating an International Criminal Court;
- Creating a new Economic Security Council;
- Creating a new People's Assembly, and
- Regulating multinational corporations.[16]

None of the recommendations in the report are new as all have been proposed in a variety of documents for decades.[17] This report, however, was the first time the Comprehensive Plan for Global Governance had been published with the approval and funding of the UN, even being supported by the then Secretary-General Boutros Boutros-Ghali.[18] As you can see, some of these changes like the ICC have already taken place.

In order for the UN to truly govern the world, it would have to undergo some drastic changes. Some of these changes include eliminating the veto system, establishing new organizations and forming a UN army/militia. Changes such as these will probably take a very long time and will come in very small steps. Additionally, more than likely the world government will not be created through forming these new organizations from the beginning. At present, the UN has a very solid base for further development of global governance ideas, but to form a world government would mean redefining the role of all international organizations. This would include organizations in the UN system, but also the regional organizations and even the international NGO sector (e.g. International Red Cross).

Indian philosopher P. R. Sarkar wrote that part of the problem is the fear of losing local leadership. Sarkar believes that the strength, by and large, of geo-political and social sentiments (casteism, racism, nationalism) will continue to fade over time.[19] Sarkar also argues for two houses/parliaments. He suggests that the first would have representatives based on population and the second on nation. Both houses would have to ratify decisions. It states, "Initially, the world government will be legislative but only in certain areas. This will eventually expand. But world governance must be based on more than a theory of collective security; it must be fundamentally cultural, humanitarian, a belief that local cultures combined can create a new global human culture and retain their own individual aesthetics."[20] Whatever the case may be, the process for Sarkar must be incremental.

Dr. Sohail Inayatullah has summarized the main reform-oriented trends. They are:

- Transform Security Council (and make it more representative of real power).
- Change structure of power within UN (between the SG, the UNGA, and the SC as well as UN bureaucracy) by increasing the power of the SG, or transforming the power of the SC or making the UNGA more representative.
- Democratize UN (by better representation of aspirations of the world). This could mean not only within statist forms by, for example, diluting the veto, but also by allowing for some type of role for NGOs beyond consultative status.
- Make UN more accountable (treat UN as a business instead of a large bureaucracy functioning through political, state-level patronage) and thus more responsive.
- Redesign the UN—two houses, four houses, regional associations or some other design structure.
- Rethink peacekeeping—creating a military with soldiers not from nations but a professional standing army.
- Popularize UN—create a house of NGOs or social movements that reflect the values of the women's movement, ecology, positive peace, spiritual transformation, social justice and sustainability. Develop an annual State of Humanity address.
- Strengthen UN—more powers, more military powers, more peacekeeping, more development, and more funding for the UN.
- Become a World Government—with legislative power initially and eventually executive and judicial powers; also deny national sovereignty when necessary.[21]

Former UN Secretary-General Kofi Annan believes the UN is at the proverbial "fork in the road". One path leads to a strengthened and improved UN made capable of resolving our serious global problems while the other path is to do nothing which will lead to eventual disaster for humanity. The decision is not only up to the rich and powerful, but it is up to all the citizens of the world "to become involved and apply sufficient collective persuasion and pressure to ensure their governments act responsibly so the UN takes the correct path in the road that lies ahead."[22] What former U.S. President John F. Kennedy stated in his 1961 inaugural address must still be true today: "The United Nations, our last best hope in an age where the instruments of war have far outpaced the instruments of peace ..."[23]

## Important words on global governance

Establishing a world government will not be easy and it will not happen overnight. However, more and more people including scholars and politicians are realizing that it is inevitable. In the future, whatever route is taken, federalism is the most likely answer. It has already proved the most successful of all political experiments. Many organizations, as for example the World Federalist Association, have advocated it for decades as the basis for global government.[24]

The regional unions throughout the world such as the EU and international organizations including the UN are helping to push ourselves forward. With these organizations and international documents in place, the road has already been paved for world government. The real question is whether or not we will we stay on that path. As for humanity as a whole, if federally united, we won't really be so very far from those much earlier ancestors, the ones huddled around that primeval fire by the cave entrance; it's just that by then the whole world will be our valley.[25]

Backdropped by Earth, this full view of the International Space Station was photographed from the Space Shuttle *Discovery* during the STS-114 Return to Flight mission, following the undocking of the two spacecraft.

On the next few pages are two very important and enlightening documents from two of the most influential people of the 20th century. Are their words of wisdom just a fantasy or can we one day make it a reality?

Mont Tremblant, Canada, October 8, 1999, Speech at the International Conference of Federalism[26]

**Confrontation or cooperation**
By Bill Clinton

*Preliminary Remarks*
Prime Minister Chretien; to the Prime Minister of St. Kitts and Nevis, Linda Douglas; Premier Bouchard; Co-chairs of this conference, Bob Rae and Henning Voscherau; to distinguished visitors, governors—I think the Lt. Governor of South Dakota, Carole Hillard, is here. And to all of you. I think it is quite an interesting thing that we have this impressive array of people to come to a conference on federalism, a topic that probably ten or 20 years ago would have been viewed as a substitute for a sleeping pill. (*Laughter*)

But in the aftermath of the conflicts in the former Yugoslavia; the interesting debates—at least I can say this from the point of view as your neighbor—that has gone on in Quebec; the deepening, troubling efforts to reconcile different tribes who occupy nations with boundaries they did not draw in Africa; and any number of other issues, this topic of federalism has become very, very important.

It is fitting that the first global conference would be held here in North America, because federalism began here—a founding principle forged in the crucible of revolution, enshrined in the Constitution of the United States, shared today by all three nations on our continent, as I'm sure President Zedillo said.

It is also especially fitting that this conference be held in Canada. A land larger than China, spanning five time zones and ten distinct provinces, it has shown the world how people of different cultures and languages can live in peace, prosperity and mutual respect.

In the United States, we have valued our relationship with a strong and united Canada. We look to you; we learn from you. The partnership you have built between people of diverse backgrounds and governments at all level is what this conference is about—and, ultimately, what democracy must be about, as people all over the world move around more, mix with each other more, live in close proximity more.

Today, I would like to talk briefly about the ways we in the United States are working to renew and redefine federalism for the 21st century; then, how I see the whole concept of federalism emerging internationally; and finally, how we—how I think, anyway—we should judge the competing claims of federalism and independence in different contexts around the world.

*American Federalism at the Dawn of the 21st Century*

First let me say we are 84 days, now, from a new century and a new millennium. The currents of change in how we work and live and relate to each other, and relate to people far across the world, are changing very rapidly.

President Franklin Roosevelt once said that new conditions impose new requirements upon government and those who conduct government. We know this to be the case not only in the United States and Canada, Great Britain and Germany, Italy and France, Mexico and Brazil, but indeed, in all the countries of the world. But in all these places there is a federalist system of some form or another. We look for ways to imbue old values with new life, and old institutions with new meaning.

In 1992, when I ran for President, there was a growing sense in the United States that the compact between the people and their government, and between the states and the federal government was in severe disrepair. This was driven largely by the fact that our federal government had quadrupled the national debt in 12 years, and that had led to enormous interest rates, slow growth and grave difficulties on all the states of our land which they were powerless to overcome.

So when the Vice-President and I ran for national office, we had no debate from people who said, look, this is a national priority and you have to deal with it. But we talked a lot to governors and others about the necessity to create again what our founding fathers called the laboratories of democracy. We, frankly, admitted that no one knew all the answers to America's large welfare caseload, to America's enormous crime rate, to America's incredible diversity of children and challenges in our schools. And so we said we would try to give new direction to the nation and deal with plainly national problems, but we would also try to build a new partnership that would make all of our states feel more a part of our union and more empowered and determined in their own destiny.

Now, people develop this federalist systems for different reasons. It came naturally to the United States because Great Britain set up colonies here as separate entities. And the states of our country actually created the national government. So we always had a sense that there were some things the states were supposed to do and some things the federal government were supposed to do.

Our founding fathers gave us some indication in the Constitution, but the history of the United States Supreme Court is full of cases trying to resolve the whole question of what is the role and the power of the states as opposed to what is the role and the power of the national government in ever-new circumstances.

There are different examples elsewhere. For example, in the former Yugoslavia when it existed before, federalism was at least set up to give the appearance that all the different ethnic groups could be fairly treated and could have their voices heard. So in 1992, it appeared that the major crisis in federalism was that the states had been disempowered from doing their jobs because the national economy was so weak and the fabric of the national society was fraying in America. But underneath that I knew that once we began to build things again we would have to resolve some very substantial questions, some of which may be present in your countries, as well.

As we set about to work, the Vice-President and I, in an effort that I put him in charge of, made an attempt to redefine the mission of the federal government. And we told the people of the United States that we actually thought the federal government was too large in size, that it should be smaller, but more active; and that we should do more in partnerships with state and local governments and the private sector, with the ultimate goal of empowering the American people to solve their own problems in whatever unit was most appropriate, whether it was an individual citizen, the family, the community, the state or the nation.

And we have worked at that quite steadily. Like Canada, we turned our deficit around and produced a surplus. We also shrank the size of the federal government. The size of the United States federal government today is the same as it was in 1962, when John Kennedy was President, and our country was much, much smaller.

In the economic expansion we have been enjoying since 1993, the overwhelming majority of the jobs that were created were created in the private sector. It's the largest percentage of private sector job creation of any economic expansion in America since the end of World War II. Meanwhile, many of our state and local governments have continued to grow in size, to meet the day-to-day demands of a lot of the domestic issues that we face in our country. And I think that is a good thing.

In addition to shrinking the size of government, we've tried to empower the states to make more of their own decisions. For example, the Department of Education has gotten rid of two-thirds of the rules that it imposed on states and school districts when I became President. Instead, we say, here are our national objectives; here is the money you can have; you have to make a report on the progress at meeting these national objectives, but we're not going to tell you how to do it anymore. And it's amazing what you can do if you get people to buy into national objectives with which they agree, and you stop trying to micromanage every instance of their lives and their daily activities. So we found some good success there.

We've also tried to give the states just blanket freedom to try more new ideas in areas where we think we don't have all the answer now, from health policy to welfare reform, to education to fighting crime.

We have always felt—this has been easy in the United States, though, compared to a lot of places because we've had this history of believing from the time of our founders that the national government would never have all the answers, and that the states should be seen as our friends and our partners because they could be laboratories for democracy. They could always be out there pushing the envelope of change. And certain things would be possible politically in some places that would not be possible in others.

And we have been very well served by that. It has encouraged a lot of innovation and experimentation. Here is the problem we have with the basic business of government and federalism today. In the 21st-century world, when we find an answer to a problem, very often we don't have time to wait for every state to agree that that's the answer. So we try to jump-start the federalist experience by looking for ideas that are working and then embodying them in federal legislation and giving all the states the funds and other support they need to do it.

Why do we do this? Well, let me give you one example. In 1787, in the United States, the founding fathers declared that all the new territories would have to set aside land for public schools, and then gave the responsibility for public education to the states. Now, then, in the next few years, a handful of states mandated education. But it took more than 100 years for all of our states to mandate free public education for all of our children. That was 19th-century pace of change. It's inadequate in the 21st century.

So I have tried to do what I did as a governor—if something is working in a state I try to steal it, put it into federal law, and at least give all the states the opportunity and the money necessary to influence the same change. But it's very, very important.

Since our Ambassador is a native of Georgia, I'll give you one example. One of my goals is to make universal access to colleges and universities in America, and we now have something called the HOPE Scholarship, modeled on Ambassador Giffin's home state program, which gives all students enough of a tax subsidy to afford at least the first two years of college in America, because we found in a census that no matter where you come from in the United States, people with at least two years of education after high school tended to get jobs where their incomes grew and they did better. People with less than that tended to get jobs where their incomes stayed level or declined in the global economy.

Now, we've also tried to make dealing with Washington less of a problem. We've ended something that was very controversial, at least prospectively, called unfunded mandates, where the federal government would tell the states they had to do something and give them about 5% of the money it cost to do it. That, I think, is a problem in every national federal system. We continue to give the states greater freedom and flexibility. And this summer I signed a new executive order on federalism which would reaffirm in very specific ways how we would work in partnership and greater consultation with state and local officials.

Federalism is not a fixed system; it, by definition, has to be an evolving system. For more than 200 years the pendulum of power has swung back and forth one way or the other. And I do want to say—for those of you who may be looking outside in thinking the Americans could never understand our problems, they don't have any problems like this—it is true that, by and large, in our state units we don't have people who are of just one racial or ethnic or religious groups. But, to be sure, we have some of that. I'll give you one example that we're dealing with today.

The United States Supreme Court has to decide a case from the state of Hawaii in which the state has given native Hawaiians, Pacific Islanders, the right to vote in a certain kind of election, and only native Hawaiians. And someone in Hawaii has sued them, saying that violates the equal protection clause of the United States Constitution. We disagree because of the purpose of the election.

But you can see this is a federalist issue. We basically said the national government would give that to the states, the states want to do it this way; then a citizen says no, you can't do that under national law.

Another example that causes us a lot of problems in the West—what happens when the federal government actually owns a lot of the land and the resources of a state? The national government is most unpopular in America in states like Wyoming, or Idaho, where there aren't very many people, there's a lot of natural resources. Cattlemen, ranchers have to use land that belongs to the federal government, and we feel that we have to protect the land for multiple uses, including environmental preservation as well as grazing or mining or whatever. And so it's an impossible situation.

It's very funny—in these states, when we started, the federal government was most popular in the areas where we owned most of the land, because we built dams and channeled rivers, and provided land for people to graze their cattle. And within 50 years, the federal government has become the most unpopular thing imaginable. Now, I used to go to Wyoming on vacation just to listen to people tell me how terrible the job I had was. (*Laughter*) But it's a problem we have to face.

And let me say one other thing I think might be interesting to you is that the Democratic Party and the Republican Party in the United States tend to have different ideas about federalism depending on what the issue is—which is why it's always good to have a dynamic system.

For example, we Democrats, once we find something working at the local level that advances our social policy, or our economic policy, we want to at least make it a national option, if not a national mandate. When I became President, crime was going up, but there were cities where crime was going down. I went there and found out why it was going down. And it was obvious to me we didn't have enough police officers preventing crime in the first place, so I said we're going to create 100,000 police at the national level, and give them to the cities.

The conservatives were against that. They said, you're interfering with states and local rights, telling them how to fight crime. Of course, I wasn't; I was giving them police, they didn't have to take them if they didn't want them. (*Laughter*) And it turned out they liked it quite well and we have the lowest crime rate in 26 years. But there was a genuine federalism dispute.

Now we're having the same dispute over teachers. We have the largest number of children in our schools in history; lots of evidence that smaller classes in the early grades yield permanent learning gains to children. So I said, now let's put 100,000 teachers out there. And they say I'm trying to impose this terrible burden on state and local government, sticking my nose in where it doesn't belong.

On the other hand, in the whole history of the country, personal injury law, including economic injuries, commercial law has always been the province of state and local government except for things like securities, stocks, bonds, things that required a national securities market. But many people in the Republican Party believe that since there is essentially a national economy and an international economic environment, that we should take away from the states all their states' rights when it comes to determining the rules under which people can sue businesses. And they really believe it.

And I have agreed with them as it applies to securities litigation because we need a national securities market. But I have disagreed with them as it applies to other areas of tort reform where they think it's a bad thing that there are state rights.

And I say this not to attack the other party, but only to illustrate to all of you that in whatever context you operate, there will always be differences of opinion about what should be done nationally and what should be done at the state level. That cannot be eliminated. The purpose of federalism, it seems to me, is, number one, to take account of the genuinely local feelings which may be, in the United States, a result of economic activities and ties to the land and history; or it may be in another country the result of the general segregation of people of various racial, ethnic or religious groups into the provinces in the federal system. So the first process is to give people a sense of their identity and autonomy. And then you have to really try to make good decisions so that the system works. I mean, in the end, all these systems only have integrity if the allocation of decision-making authority really produces results that people like living with, so they feel that they can go forward.

*Global Perspective on Federalism: an Overview*

Now, let me just discuss a minute what is sort of the underlying tension here that you see all across the world, which is, what is the answer to the fact that on the edge of a new millennium—where we would prefer to talk about the Internet, and the decoding of the human gene, and the discovery of billions of new galaxies in outer space—those of us in politics have to spend so much time talking about the most primitive slaughter of people based on their ethnic or racial or religious differences.

The great irony of the turning of the millennium is that we have more modern options for technology and economic advance than ever before, but our major threat is the most primitive human failing—the fear of the other, and the sense that we can only breathe and function and matter if we are somehow free of the necessity to associate with and deal with, and maybe even under certain circumstances subordinate our own opinions to, the feelings of them—people who are different from us; a different race, a different religion, a different tribe.

And there is no answer to this that is easy. But let me just ask you to look in the context of the former Yugoslavia, where we are trying to preserve a Bosnian state—Prime Minister Chretien and I, and our friends—which serves Croatians and Muslims, after four years of horrible slaughter, until we stopped it in 1995. Or in Kosovo, where we're exploring whether Kosovo can continue to be an autonomous part of Serbia, notwithstanding the fact that the Serbs ran all of them out of the country and we had to take them back. Why did all this happen? Partly because it was an artificially imposed federalism. Marshal Tito was a very smart man who basically said, I'm going to create federalism out of my own head. I'm going to mandate the participation of all these groups in government. And I'm going to forbid my government from talking about ethnic superiority, or oppression, or problems. He wouldn't even let them discuss the kind of ethnic tensions that are just part of the daily life in most societies in this world. And it all worked until he died. And then it slowly began to unravel.

*How to Judge Independence Claims in the Current Context*

So one of the reasons you have all these people clamoring for the independence of ever smaller groups is that they had a kind of phony federalism imposed from the top down. So the first lesson I draw from this is every federalist system in the world today—a world in which information is widely shared, economic possibilities are at least—always, to some extent, based on global forces, certainly in terms of how much money you can get into a country—the federalism must be real. There must be some real sense of shared authority. And people must know they have some real range of autonomy for decisions. And it must more or less correspond to what they perceive they need to accomplish.

On the other hand, it seems to me that the suggestion that a people of a given ethnic group or tribal group or religious group can only have a meaningful communal existence if they are an independent nation—not if there is no oppression, not if they have genuine autonomy, but they must be actually independent—is a questionable assertion in a global economy where cooperation pays greater benefits in every area of life than destructive competition.

Consider, for example, the most autonomous societies on Earth. Arguably, the tribes still living in the rainforests on the Island of New Guinea. There are 6,000 languages still existing in the world today. And 1,000 of them can be found in Papua New Guinea and Irian Jaya where tribes living ten, 20 miles from one another have complete self-determination. Would you like that?

On the other hand, consider the terrible problems of so many African peoples where they're saddled with national borders drawn for them at the Conference of Berlin in 1885, that took no reasonable account of the allocation of the tribes on certain lands and the history of their grazing, their farming, their moving.

So how to work it out? There is no answer. We have to provide a framework in which people can work it out. But the only point I want to make to you today—I don't want to beat this to death, because we could stay here for a week discussing this—is that at the end of World War I, the European powers I think—and America sort of withdrew, so we have to share part of the blame—but our record is not exactly spotless in how we went about carving up, for example, the aftermath of the Ottoman Empire. And so we have spent much of the 20th century trying to reconcile President Woodrow Wilson's belief that different nations had the right to be free—nations being people with a common consciousness—had a right to be a state.

And the practical knowledge that we all have that if every racial and ethnic and religious group that occupies a significant piece of land not occupied by others became a separate nation—we might have 800 countries in the world and have a very difficult time having a functioning economy or a functioning global polity. Maybe we would have 8,000—how low can you go?

So that doesn't answer any specific question. It just means that I think when a people thinks it should be independent in order to have a meaningful political existence, serious questions should be asked: Is there an abuse of human rights? Is there a way people can get along if they come from different heritages? Are minority rights, as well majority rights, respected? What is in the long-term economic and security interests of our people? How are we going to cooperate with our neighbors? Will it be better or worse if we are independent, or if we have a federalist system?

I personally believe that you will see more federalism rather than less in the years ahead, and I offer as Exhibit A the European Union. It's really a new form of federalism, where the states—in this case, the nations of Europe—are far more important and powerful than the federal government, but they are giving enough functions over to the federal government to sort of reinforce their mutual interest in an integrated economy, and in some integrated political circumstances.

In a way, we become more of a federalist world when the United Nations takes a more active role in stopping genocide in places in which it was not involved, and we recognize mutual responsibilities to contribute and pay for those things.

So I believe we will be looking for ways, over and over and over again—the Prime Minister and I have endorsed the Free Trade Area of the Americas. We'll be looking for ways to integrate our operations for mutual interest, without giving up our sovereignty. And where there are dissatisfied groups in sections of countries, we should be looking for ways to satisfy anxieties and legitimate complaints without disintegration, I believe.

That's not to say that East Timor was wrong. If you look at what the people in East Timor had been through; if you look at the colonial heritage there; if you look at the fact that the Indonesians offered them a vote, they took it, and nearly 80% of them voted for independence—it seems that was the right decision.

But let us never be under the illusion that those people are going to have an easy path. Assuming that those of us who are trying to support them help them; assuming we can stop all the pro-integrationist militias from oppressing the people, and we can get all the East Timorese back home, and they'll all be safe—there will still be less than a million of them, with a per-capita income among the poorest in the world, struggling to make a living for their children in an environment that is not exactly hospitable.

Now, does that mean they were wrong? No. Under the circumstances they faced, they probably made the only decision they could have. But wouldn't it have been better if they could have found their religious, their cultural, their ethnic and their economic footing—and genuine self-government—in the framework of a larger entity which would also have supported them economically? And reinforced their security instead of undermining it? It didn't happen; it's too bad.

But I say this because—I don't think there are any general rules, but I think that, at the end of World War I, when President Wilson spoke, there was a general assumption, because we were seeing empires break up—the Ottoman Empire, the Austro-Hungarian Empire; there was the memory of the Russian Empire; British colonialism was still alive in Africa, and so was French colonialism—at that time, we all assumed, and the rhetoric of the time imposed the idea that the only way for people to feel any sovereignty or meaning was if they were independent.

And I think we've spent a lot of the 20th century minimizing the prospects of federalism. We all have recoiled, now, so much at the abuse of people because of their tribal, racial and religious characteristics, that we tend immediately to think that the only answer is independence.

*Conclusion*

But we must think of how we will live after the shooting stops, after the smoke clears, over the long run. And I can only say this, in closing: I think the United States and Canada are among the most fortunate countries in the world because we have such diversity—sometimes concentrated, like the Inuits in the north; sometimes widely dispersed within a certain area, like the diversity of Vancouver.

We are fortunate—because life is more interesting and fun when there are different people who look differently and think differently, and find their way to God differently. It's an interesting time. And because we all have to grow and learn when we confront people who are different than we are, and instead of looking at them in fear and hatred and dehumanization, we look at them and see a mirror of ourselves and our common humanity.

I think if we will keep this in mind—what is most likely to advance our common humanity in a smaller world; and what is the arrangement of government most likely to give us the best of all worlds—the integrity we need, the self-government we need, the self-advancement we need—without pretending that we can cut all the cords that bind us to the rest of humanity—I think more and more and more people will say, this federalism, it's not such a bad idea.

Thank you very much.

United Nations World New York, October 1947, pp. 13-14 Open Letter[27]

**To the General Assembly of the United Nations**
By Albert Einstein

As I see it, this is the way for the nations of the world to break the vicious circle which threatens the continued existence of mankind, as no other situation in human history has ever done.

We are caught in a situation in which every citizen of every country, his children, and his life's work, are threatened by the terrible insecurity which reigns in our world today. The progress of technological development has not increased the stability and the welfare of humanity. Because of our inability to solve the problem of international organization, it has actually contributed to the dangers which threaten peace and the very existence of mankind.

The delegates of 55 governments, meeting in the second General Assembly of the United Nations, undoubtedly will be aware of the fact that during the last two years—since the victory over the Axis powers—no appreciable progress has been made either toward the prevention of war or toward agreement in specific fields such as control of atomic energy and economic cooperation in the reconstruction of war-devastated areas.

The United Nations cannot be blamed for these failures. No international organization can be stronger than the constitutional powers given it, or than its component parts want it to be. As a matter of fact, the United Nations is an extremely important and useful institution provided the peoples and governments of the world realize that it is merely a transitional system toward the final goal, which is the establishment of a supranational authority vested with sufficient legislative and executive powers to keep the peace. The present impact lies in the fact that there is no sufficient, reliable supranational authority. Thus the responsible leaders of all governments are obliged to act on the assumption of eventual war. Every step motivated by that assumption contributes to the general fear and distrust and hastens the final catastrophe. However strong national armaments may be, they do not create military security for any nation nor do they guarantee the maintenance of peace.

There can never be complete agreement on international control and the administration of atomic energy or on general disarmament until there is a modification of the traditional concept of national sovereignty. For as long as atomic energy and armaments are considered a vital part of national security no nation will give more than lip service to international treaties. Security is indivisible. It can be reached only when necessary guarantees of law and enforcement obtain everywhere, so that military security is no longer the problem of any single state. There is no compromise possible between preparation for war, on the one hand, and preparation of a world society based on law and order on the other.

Every citizen must make up his mind. If he accepts the premise of war, he must reconcile himself to the maintenance of troops in strategic areas like Austria and Korea; to the sending of troops to Greece and Bulgaria; to the accumulation of stockpiles of uranium by whatever means; to universal military training, to the progressive limitation of civil liberties. Above all, he must endure the consequences of military secrecy which is one of the worst scourges of our time and one of the greatest obstacles to cultural betterment.

If on the other hand every citizen realizes that the only guarantee for security and peace in this atomic age is the constant development of a supranational government, then he will do everything in his power to strengthen the United Nations. It seems to me that every reasonable and responsible citizen in the world must know where his choice lies.

Yet the world at large finds itself in a vicious circle since the United Nations powers seem to be incapable of making up their minds on this score. The Eastern and Western blocs each attempt frantically to strengthen their respective power positions. Universal military training, Russian troops in Eastern Europe, United States control over the Pacific Islands, even the stiffening colonial policies of the Netherlands, Great Britain and France, atomic and military secrecy—are all part of the old familiar jockeying for position.

The time has come for the United Nations to strengthen its moral authority by bold decisions. First, the authority of the General Assembly must be increased so that the Security Council as well as all other bodies of the United Nations will be subordinated to it. As long as there is a conflict of authority between the Assembly and the Security Council, the effectiveness of the whole institution will remain necessarily impaired.

Second, the method of representation at the United Nations should be considerably modified. The present method of selection by government appointment does not leave any real freedom to the appointee. Furthermore, selection by governments cannot give the peoples of the world the feeling of being fairly and proportionately represented. The moral authority of the United Nations would be considerably enhanced if the delegates were elected directly by the people. Were they responsible to an electorate, they would have much more freedom to follow their consciences. Thus we could hope for more statesmen and fewer diplomats.

Third, the General Assembly should remain in session throughout the critical period of transition. By staying constantly on the job, the Assembly could fulfill two major tasks: first it could take the initiative toward the establishment of a supranational order; second, it could take quick and effective steps in all those danger areas (such as currently exist on the Greek border) where peace is threatened.

The Assembly, in view of these high tasks, should not delegate its powers to the Security Council, especially while that body is paralyzed by the shortcomings of the veto provisions. As the only body competent to take the initiative boldly and resolutely, the

United Nations must act with utmost speed to create the necessary conditions for international security by laying the foundations for a real world government.

Of course there will be opposition. It is by no means certain that the USSR—which is often represented as the main antagonist to the idea of world government—would maintain its opposition if an equitable offer providing for real security were made. Even assuming that Russia is now opposed to the idea of world government, once she becomes convinced that world government is nonetheless in the making her whole attitude may change. She may then insist on only the necessary guarantees of equality before the law so as to avoid finding herself in perennial minority as in the present Security Council.

Nevertheless, we must assume that despite all efforts Russia and her allies may still find it advisable to stay out of such a world government. In that case, and only after all efforts have been made in utmost sincerity to obtain the cooperation of Russia and her allies—the other countries would have to proceed alone. It is of the utmost importance that this partial world government be very strong, comprising at least two-thirds of the major industrial and economic areas of the world. Such strength in itself would make it possible for the partial world government to abandon military secrecy and all the other practises born of insecurity.

Such a partial world government should make it clear from the beginning that its doors remain wide open to any non-member—particularly Russia—for participation on the basis of complete equality. In my opinion, the partial world government should accept the presence of observers from non-member governments at all its meetings and constitutional conventions.

In order to achieve the final aim—which is one world, and not two hostile worlds—such a partial world government must never act as an alliance against the rest of the world. The only real step toward world government is world government itself.

In a world government the ideological differences between the various component parts are of no grave consequence. I am convinced that the present difficulties between the USA and the USSR are not due primarily to ideological differences. Of course, these ideological differences are a contributing element in an already serious tension. But I am convinced that even if the USA and Russia were both capitalist countries—or communist, or monarchist, for that matter—their rivalries, conflicting interests and jealousies would result in strains similar to those existing between the two countries today.

The United Nations now and world government eventually must serve one single goal: the guarantee of the security, tranquility, and the welfare of all mankind.

Signed by A. Einstein

# Chapter 10

# What does the future hold?

"Only within the moment of time represented by the present century has one species—man—acquired significant power to alter the nature of his world."
—Rachel Carson, writer, scientist, ecologist

Artist's concept of astronauts at a moon base.

After reading about the problems we face as a species along with that of our planet, will anyone be willing to stand up and take action? Will the necessary changes occur to combat these crises or will the world continue down what seems to be a path of destruction?

In the very beginning of the book I stated that awareness is the key. You must be aware of what is taking place in order to make a positive change. However, throughout history we have witnessed thousands of instances where good men and women sat back and did nothing. A recent example would be in 1994 when the world took a blind eye to hundreds of thousands of people being murdered in Rwanda. Will another tragedy occur because good people fail to act? Will the human race sit back and watch while we destroy ourselves and our planet?

# 2050—a very real scenario

"The only thing necessary for the triumph of evil is for good men to do nothing." —Edmund Burke, British political writer.

It is the year 2050, and no one can believe it has come to this. Everyone is wondering how we, as a global civilization, let the world get to this point. The human population has grown to almost ten billion. This in turn has affected almost every aspect of life. The biggest visible change to all creatures inhabiting the planet has been the environment. Due to rising temperatures, mainly because of increased levels of carbon dioxide and other greenhouse gases, almost a million of Earth's land-dwelling plants and animals are now extinct. Another 25% of the species left on the planet are also at risk of extinction.

Since the ice caps continued to melt, the ocean temperature and level rose. As a result, coastlines have been devastated everywhere. The rising ocean level eventually made its way into the drinking water supplies in many communities and countries, especially along the east coast of the United States.

The air temperature has also risen dramatically and the ozone layer is almost non-existent. This has brought about horrific changes to humans with one common example being more heat-related deaths and illnesses. Furthermore, worsening air quality has led to a rise in cardiovascular and respiratory diseases. The number of people living with cancer, especially skin cancer, has skyrocketed. Infectious diseases have also become a common occurrence because of new geographic ranges and activity of disease-carrying animals, insects, and infective parasites.[1] Many countries tried to minimize exposure to these diseases through traditional methods such as spraying insecticides to control insects and other disease-bearing organisms, but conversely these pesticides led to more deaths themselves. Moreover, due to the high temperatures and the extreme rates many were killing parasites, the parasites evolved ever-increasing resistance to drugs and control methods. Now infectious diseases such as encephalitis and dengue (a severe influenza-like disease) run rampant.[2]

The rising air and water temperatures also had a huge effect on the marine ecosystem. In the earlier 21st century, marine phytoplankton and zooplankton sheltered a dormant form of cholera when pH, temperature, salinity, and nutrient levels are insufficient to sustain the infectious form.[3] But when waters warmed up, the nutrient levels increased, and cholera can become infectious again, spreading everywhere. The higher water temperatures have also led to an increase in diseases transmitted from fish and shellfish. The higher surface temperatures have stimulated the growth of certain species of algae, in particular, toxic 'red tides'.[4] These algae are then consumed by fish and shellfish, which have resulted in the poisoning of thousands of humans. Moreover, the rising ocean level eventually made its way into the drinking

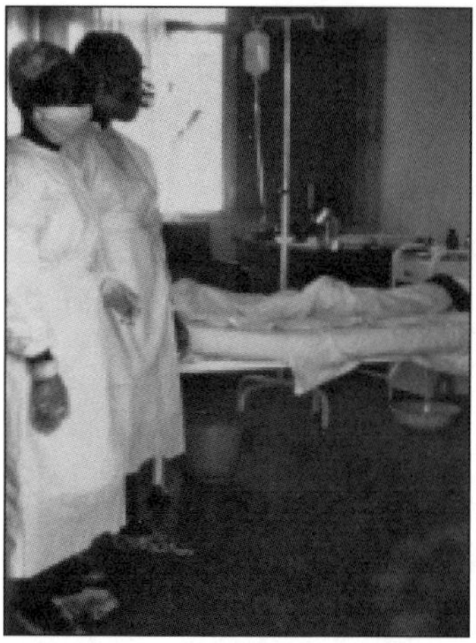

Two nurses standing in front of Ebola case #3 in 1976 (DRC).

water supplies in many communities and countries, especially along the east coast of the United States.

Unfortunately the less developed countries have been the hardest hit. Almost the entire Southern Hemisphere has been overrun with famine and disease. To make things worse, the whole continent of Africa is a war zone, as millions upon millions fight for access to scarce food and water. In addition, thousands are dying of older, more basic diseases but also due to newer more sophisticated viruses that have mutated to ensure their own survival. The world powers, or the 'West', have not intervened.

After decades of fighting, the Western countries continue their so-called 'struggle against Islam', as most know they want access to the only remaining oil supplies on the planet, all of which are located within the borders of the remaining Islamic states. What started off 47 years earlier with Iraq and then Lebanon has overflowed to all nations in the region. The U.S. and others have now resorted to using smaller tactical nuclear weapons to limit the deaths of their military personnel, as their forces are stretched to the limit across the entire region. Besides the death and destruction caused by the nuclear blasts, hundreds are being killed daily by smaller attacks such as car bombs and air-to-surface missiles. The Israel/Palestine region has been fully destroyed and is now completely deserted.

Terrorism also runs rampant throughout the world. On any given day you will hear the news of another terrorist bombing that has occurred in Asia, Europe or the Americas. Unlike the terrorists of the old days, who used civilian aircraft as flying bombs, they now have in their possession both chemical and biological weapons. The situation has become so dire, that almost all Americans have their own personal biological/chemical protective suit.

Not only has this war for oil led to various conflicts and violence around the world, the price of oil and eventually its lack of supply has devastated the economies of most of the countries around the world. It began when the price of oil reached US$100 dollars a barrel. Instead of the price stabilizing it continued to skyrocket higher and higher. Businesses, among others, were unable to cope with the

A B-2A bomber releases a test version of the new B61-11 gravity bomb over the Tonopah Test Range in Nevada, November 20, 1996.

situation and the economies started to take a nosedive. Governments tried to implement the use of alternative energy sources, but they simply did not have the time or the resources.

Unfortunately this is only the half of it. Another record-breaking hurricane brews in the Pacific. London is slowing going underwater. The list of devastations seems endless. Could we have done something differently?

## 2050—an alternative view

"I like to believe that people in the long run are going to do more to promote peace than our governments. Indeed, I think that people want peace so much that one of these days governments had better get out of the way and let them have it."—Dwight D. Eisenhower, former U.S. president

It is the year 2050 and some cannot believe the accomplishments the world has seen in the past four decades. From landing on Mars to finding a vaccine for HIV/AIDS, the world could never be better. When looking back at all the achievements, all would agree that the vital turning point in history was the day the planet and its people came together as one.

Following is a copy of the newspaper article from that historical day and event.

**The *Planet's Post***

People of earth come together to form world government

By Scott Firsing, *Planet's Post* Writer *Fri July 4, 2026; 3:45 PM ET*

> NEW YORK—At a press conference today at the United Nations, on the 250th anniversary of the United States of America, a diverse coalition of world leaders, celebrities, Nobel Laureates, and NGOs called for "a just and democratic New World Order."[5]
>
> After years of debate and the witnessing of several disasters, the ambitious plan, spelled out in a 423-page document titled 'One World: A Plan for Just and Democratic Global Governance', was endorsed at the conference by a who's who of political, religious, cultural, scientific, and corporate leaders. Some of the people in attendance included the leaders of all countries, the Catholic Pope

and his Holiness the Dalai Lama.
"We are living in a global village that is victimized daily by war and the constant threat of war and by an international state that borders on anarchy. Like the founding fathers of the United States of America, we must create 'a United States of the world', with a 21st-century constitution to safeguard the planet and all its creatures," said the current Secretary-General of the United Nations.[6]
"The greatest hope for a solution to humanity's gravest problems—war and terrorism, poverty, tyranny, the destruction of the environment—lies in the creation of a democratic world government," said the Pope.
The document, co-authored by 33 prominent experts on government, international law, defense, economics, the environment, and related subjects was a joint project of states, IGOs and over 200 NGOs.
The Secretary General (SG) closed the press conference with the following words of famous political philosopher and drafter of America's Declaration of Independence, Thomas Jefferson. "The care of human life and happiness, and not their destruction, is the first and only object of good government." The SG concluded by saying, "Today we come together as a global civilization to form a world government which will ensure that the human race will never fall victim to destruction. From this day forward, we must not dwell on the horrors that have occurred in the past; rather we shall always keep our eyes focused on the bright future that lies ahead."

Tad Daley wrote, "When I speak I tell the story of the crew of the *Apollo 8* spacecraft, floating over the mountains of the far side of the moon, and then seeing—for the first time with human eyes—our single, borderless, breathtaking planet, suspended among the blazing stars. Our singular and paramount loyalty, to our single and lonely planet, can be the single greatest engine taking our movement from what we now are, to what we can become."[7]

A picture of pure beauty. Sodwana Bay, South Africa, in the early morning.

That historic day in 2026 started that engine. Since that day, everything began to get constantly better. More resources were put into fields such as global health, education, the environment, and science. New discoveries and inventions were constantly being published and implemented. Crime and terrorism seemed almost to disappear. It was as if billions of people finally opened their eyes and saw what needed to be seen, almost as if they were part of the *Apollo 8* mission looking down on Earth. Below are a few of the more important triumphs from 2026 to 2050 (separated by their particular field):

- Health: Thanks to the previous research done by organizations like the Bill and Melinda Gates Foundation, the world was able to eradicate dozens of diseases/viruses from the face of the Earth. A cure was also discovered for HIV/AIDS, Alzheimer's and schizophrenia.
- Education: Every child is now guaranteed an education. The new government has established schools throughout the world and has trained all the teachers they need.
- Environment: The world began the slow transition to alternative sources of energy, like solar and wind power. Now the world is no longer reliant on oil, as it is only used for a few specialized purposes. This change also reduced the greenhouse gases in the atmosphere and the ozone layer is slowly beginning to replenish.
- Security: Due to the new government setting up a global defence force with rapid deployment capability, war has become a thing of the past. In addition, the global police force has helped curb crime throughout the world.
- Economy: The world had finally achieved a global economy that appeared to be environmentally sustainable. Technologies such as biotechnology, nanotechnology, closed-environment agriculture and others, now help feed the world.
- Space: Starting with NASA's early 21st century goal of returning man to the moon, there are now moon bases located on all sections of the lunar surface. Additionally, more lunar houses are going up every year as people have begun making the move to live on the moon permanently. Space tourism is also the new 'in' thing. Moreover, in the 2030s man finally set foot on the red planet Mars.

As horrific as the first scenario sounds and how pleasant the second scenario would be, the plausible guess is the world in 2050 would be somewhere in between the two. The problem with predicting the future is just that, it is unpredictable. These two scenarios are just a glimpse into two possible outcomes out of thousands of possibilities, all of which will stem from the actions we take today. The main point here is not to predict the future, but to change it. Will the generation of today's youth make that decision to change the world for the better or do they simply not even care about the present situation?

## Is this the generation?

In Chapter 6 I briefly touched on the Humanitarian Youth Culture Study that was completed by Label Networks. Their company, which is a leader in global youth culture marketing intelligence and research, compared the most trend-setting and mainstream subcultures in the world. Label Networks was started in 2000, as they wanted to bridge the gap between companies trying to access new youth markets and the lack of authentic research and intelligence about this fast-changing landscape.[8]

As a refresher, the Humanitarian Youth Culture Study addressed the youth of North America on issues such as the environment and global warming, recycling, AIDS, Africa, poverty, volunteerism, politics and their purchasing practises. They asked open-ended questions such as "What change do you want to make in your life that will positively affect those around you and the world you live in?"[9] The Humanitarian Youth Culture Study was comprised of a representative sampling of 5,000 youths, ages 13-24 across North America and was conducted from May 5-June 15, 2006. Results also included actionable, macro-trend editorial summaries and answers to *why* young people responded the way that they did, including results to open-ended questions and comments, plus key cross-tabs by gender and specific age groups.[10]

Our company, "started this project for many reasons," explains Kathleen Gasperini, Senior Vice-President of Marketing and Development at Label Networks. "First, we read the Commission Report on Africa from the DATA website and One Campaign which stated that 'The Commission substantially lacks in only one area and that is the lack of a proper analysis of Western attitudes towards Africa.'"[11] She knew that her company could answer these questions in terms of the youth marketplace and that the research results and services may be of great importance and value to DATA, One.org and the Natural Resources Defense Council (NRDC) for example, because of their fresh, authentic approaches which connect the dots across cultures between local inspiration, viral trends, and global movements.[12]

This study found that the North American youth culture of today is among the most charitable and environmentally conscious demographic in the world.

They are far more aware than previous generations of what's taking place across the globe in terms of humanitarian need, poverty, AIDS, Africa, war and terrorism, and environmental issues such as global warming. As a generation growing up in the midst of increased consumer control, they not only *want* to do something about their future, they *expect* to.[13] What this means is that there's a great opportunity for reaching young people by appealing to their sense of 'good causes.' With evidence such as this, perhaps

Mya, a fourth-grade student at the North American International School, surfs the Internet looking for information on NASA and its astronauts.

this is the generation ready to change the world. Below are a few examples from the Humanitarian Study:

Would you be more willing to buy a specific brand if you knew that the brand donated a percentage of profits to non-profit organizations? 64.5% Yes; 30.9% Maybe

Generally, are you worried about the state of the environment? 49.2% Somewhat worried about the environment; 43.5% Very worried

What concerns you the most about the environment? 34.2% Gas alternatives; 31.6% Global warming; 10.1% Cleaner cities; 8.8% Cleaner oceans and rivers

What best describes how you feel about the United States in terms of the environment? 43.7% Our government is messing up with the environment and our future; 40.1% We should do more to save our environment; 4.6% I don't really care

Which best describes your perception of Africa's problems? 26.6% We need to fight the AIDS crisis there; 24.9% If we help Africa, then we are helping all of us in the end; 18.1% We need to educate more people; 17% We need to stop poverty

If you could spend your time volunteering, which project would you like to work on the most? 13.2% Youth suicide; 13.1% Big brother/Big sister; 12.1% Animal rights

Results of the Humanitarian Study include several open-ended questions. Comment examples:

"There's the war on drugs, the war on immigration, the war on AIDS, the war on war, but it is the war on earth—global warming—that's going to kill us all unless we do something."

"I feel pretty strongly about youth suicide because kids need to know that someone will be affected if they choose this path of death."

"I think we have tipped over the environmental edge and are at best hanging to a weedy growth out of the side of a cliff high above sharp rocks."

# PeaceJam

PeaceJammers with the Dalai Lama.

One organization that is committed to the youth of today is the non-profit international education program, PeaceJam. Here, Nobel Peace laureates work personally with youth to help them become engaged citizens who work for the betterment of their local communities and the world at large. According to the organization's co-founder, Dawn Engle, their "mission is to create a new generation of young people who can change the world," and she believes it is really happening.[14] Since the program's founding in 1996, more than 200,000 PeaceJammers worldwide have created and carried out more than 250,000 service-learning projects in their communities.[15] In September 2006, PeaceJam celebrated their tenth anniversary at the University of Denver. Their guests included 12 of the planet's most notable peacemakers and over 3,000 young people

How PeaceJam works is it establishes its programs by going directly to the source. This may include anything from elementary and high schools to boys and girls clubs and faith-based groups. The high-school-aged participants then create community projects, starting with a presentation of their plans to a Nobel laureate during a PeaceJam weekend. These weekends are capped at no more than 400 participants to maintain the feasibility of close personal interaction between the attendees and the laureates.[16] In the following months, the kids carry out their projects and at the end of the year, they all come together again for the PeaceJam Slam. It is here where they report on their projects, get more leadership training and hear the announcement of which laureate is coming for the next PeaceJam.

PeaceJam Youth Conference at Cape College in Cape Town, South Africa in June 2004.

This process goes on year after year. "Education, inspiration and action, and it's the same the next year: education, inspiration and action. You're creating young leaders who are empowered to go out and create change," says Engle.[17] "What is most interesting for me is the encounter with hundreds of youth from different ethnic, cultural and social backgrounds, coming together with one goal, and that is to promote human solidarity, understanding, tolerance and peace," commented Ramos-Horta at the close of the February 2006 Memphis PeaceJam. "We should not underestimate the tremendous energy and power of the youth. They can do enormous good to the world."[18]

"The transformation of kids when they work with the Nobel Peace Prize winners is incredible," says PeaceJam co-founder Ivan Suvanjieff. It gives them the courage to stand up for their convictions, to go out in the field and act upon the better instincts of helping people, and know that it's a cool thing to do."[19]

## World Future Council

"The Council is a marvelous and timely concept. If ever the planet needed to hear voices of wisdom and reason in facing humanity's common perils, it is now."—Daniel Goleman, psychologist

In many areas dealing with the global issues discussed throughout this book, we already know what needs to be done. However, most policy-makers fail to act due to political cowardice, short-sightedness and corruption.[20] In cases such as these, the World Future Council (WFC) will focus on closing this 'action gap' between what is being done and what urgently needs to be done. In the other areas, where it is not yet evident what needs to be done, additional research will be carried out to develop new, innovative proposals. The Council will consider and publicize its proposals for action at annual meetings that will be broadcast worldwide.[21]

The WFC is an initiative that aims for a global membership of no more than 100 wise thinkers, practical pioneers and young leaders. WFC believes that humanity has unprecedented powers to damage the planet and affect the wellbeing of present and future generations, yet we seem powerless to address the global and long-term impacts of our actions. That is where the WFC comes in. It challenges the short-term commercial thinking that currently has veto power over global decision-making. The Council will also work closely with national legislators from all over the world to develop step-by-step reforms and legislation to overcome the current "implementation gap".[22] The EarthAction Network with its 2,200 Partner Organizations in more than 160 countries will help build global support for the Council's recommendations.

The WFC also knows that "mere pronouncements from even an ethical group like this one will not create a better world."[23] So, the WFC will make sure its recommendations reach and influence policymakers as they allocate funds and enact and enforce both national and international laws. In addition, it will propose what is necessary and not just what seems politically possible. It will create an atmosphere of empowerment and widely shared ownership in the solutions to the problems facing humanity.

At the first Inaugural Meeting of the Founding Council and the Board of

The STS-92 Space Shuttle astronauts photographed upstate New York at sunset on October 21, 2000. Water bodies (Lake Ontario, Lake Erie, the Finger Lakes, the St. Lawrence and Niagara rivers) are highlighted by sun glint (sun reflecting off the water surface), making for a dramatic and unusual regional view. The photograph was taken looking toward the southwest from southern Canada, and captures a regional smog layer extending across central New York, western Lake Erie and Ohio, and further west. The layer of atmospheric pollution is capped by an atmospheric inversion, which is marked by the layer of clouds at the top of the photograph. The astronauts were able to document this smog event from a variety of vantage points as they orbited over the northeastern U.S. and southern Canada.

Advisors and Donors on 20-22 May 2006 in Geneva, Switzerland, Herbert Girardet summarized why the WFC is needed:

- ❑ To develop agreement on ethical standards for human action;
- ❑ To speak out as an articulate and effective lobby for future generations and responsible global citizenship;
- ❑ To set up a new international process to bridge the major implementation gaps;

- To provide the missing institutional link, working with others to increase the pressures for positive change until they can no longer be ignored.[24]

Jakob von Uexküll, President of the World Future Council Initiative, provides even more evidence of why the WFC would be beneficial. In his overview of the WFC Initiative he explains that many studies show a remarkable global convergence of common values and priorities. For example, research by the Institute for Global Ethics involving individuals of different social backgrounds and beliefs in over 20 countries on all continents found common values to be (in their order of preference): truth/honesty, love/compassion, freedom, fairness, community, tolerance, responsibility. Furthermore, the Valencia 2000 'Declaration of Human Duties and Responsibilities', drafted for UNESCO by a broad international commission, also reveals a remarkable consensus on the underlying common values.[25] However, as Jakob states: "We lack a voice which appeals and responds to our common values as planetary citizens, to our moral inner voice struggling to be heard against the round-the-clock cacophony of commercial speech. We propose the creation of such a global voice to remind us of our responsibilities as guardians of all future generations of life on earth—a 'World Future Council'. This body could provide an ethical audit on important decisions and their compatibility with our duty to assist those in danger and safeguard the world for our grandchildren. It would expose the connection between the pursuit of economic goods and the growth of social bads."[26]

Today, we know that the effects of global decisions are felt in even the most remote village. What the Council will provide is a "permanent, respected forum to ensure that these decisions are transparent and considered from the perspective of our highest values including the right of all peoples and communities to make informed choices about their level of global integration and have their choices respected."[27] Mr. von Uexküll knows that a 'business as usual' attitude is not the answer, as the growing conflicts between the proponents and opponents of economic globalization threaten to paralyze international decision-making.

In June of 2006, the WFC received the good news that the city of Hamburg, Germany, will provide the core funding for the launch and initial phase of the WFC. The full WFC inaugural conference will be held there in the second quarter of 2007. The First Mayor of the Free and Hanseatic City of Hamburg, Ole von Beust, and the prominent Hamburg entrepreneur, Dr. Michael Otto, have agreed to provide five million euros over the next three years to fund the WFC in a public-private partnership.[28] The city will contribute 50% from public funds and Dr. Otto personally 30%. The Mayor and Dr. Otto have jointly committed to raise the remaining one million euros from other donors.

Mr. von Uexküll warmly welcomed the announcement by saying, "We are delighted that Hamburg will host the Secretariat of the World Future

Council. This is a huge step forward, for which I am very grateful to everyone involved. Now begins the huge task of fulfilling the many expectations which the WFC Initiative has raised."[29] Still, additional funding will need to be raised to finance these endevours as well as to provide longer-term security and build a global organization with a presence in different parts of the world.

## Conclusion

"For those who have seen the Earth from space, and for the hundreds and perhaps thousands more who will, the experience most certainly changes your perspective. The things that we share in our world are far more valuable than those which divide us." —Donald Williams, astronaut.

Picture of the Florida Peninsula (USA). When you view photos of the world without borders, you truly see that humanity shares a common destiny.

My main goal in writing this book was to prepare humanity for a global future and its potential impact on our lives. After reading this book I trust that you now have a broader understanding of what is currently happening around us. Furthermore, I hope that by my highlighting individual and group contributions, along with this concluding reflection, you have been provided with a better knowledge of what people are willing to do for the betterment of humanity. I also hope that this book has served somewhat as a guide that will help point us all in the right direction for the sake of humankind.

We know that change is needed and that solving many of these problems relies on both the leaders of today and the young leaders of tomorrow. Moreover, we are certain that the 21st century cannot simply be a repeat of the twentieth. Just by examining the 20th-century statistics on human conflict and violence alone says more than enough. Likewise it doesn't matter how many times you look at the facts, the results are startling.

"Over 110 million people have been killed in wars in the 20th century; and there have been over 120 wars since the end of WWII in 1945. The last century invented gas chambers, total war, state-sponsored genocide and extermination camps, brainwashing, state security apparatuses, and the 'panoptic surveillance of entire populations'." It generated more victims, dead soldiers, murdered civilians, displaced minorities, torture, political prisoners, and refugees, and more dead from cold, hunger, maltreatment "than ever could have been imagined. The phenomena of violence and barbarism [some have concluded] mark the distinctive signature of the age."[30]

Will forming a world government in the upcoming decades ensure the 21st century is one of peace? I know many people in politics and other fields frown upon the idea of global governance. They question the proposal of a world government because of the weakness of human nature that could result in global tyranny amongst other things. Yet most of those same people either believe or engage in some form of global governance. This includes the UN, the ICC, Interpol and regional federations. The beginning of the 21st century has also brought with it a global ethics movement and the United Nations Millennium Declaration. Such a global ethic is a "minimal basic consensus relating to binding values, irrevocable standards, and moral attitudes which can be affirmed by all ['open' representatives of] religions and [non-dogmatic] non-believers. In particular, this global ethic also accepts a culture of non-violence and respect for life." In this culture, "persons who hold political power must work within the framework of a just order and commit themselves the most nonviolent, peaceful solutions possible."[31]

Albert Einstein stated, "The splitting of the atom has changed everything except our way of thinking, and thus we drift toward unparalleled catastrophe. Unless there is a fundamental change in [our] attitudes toward one another as well as [our] concept of the future, the world will face unprecedented disaster." Einstein also said, "that there is no salvation for civilization or even the human race, other than the creation of a world government."[32] Others agree with him as well. Benjamin Ferencz and Ken Keyes in their 1991 book, *Planethood*, state: "World government is not only a possibility, it is inevitable, because it appeals to the patriotism of people who love their national heritages so deeply that they wish to preserve them in safety for the common good."[33] I am trying to point us in the right direction, but in all actuality it is hard to imagine how we will address future domestic and international problems in a globalized world without organized global governance.

At this moment in time, wars continue. Many people still face famine, poverty, genocide, terrorism and even their denial of basic human rights. The ozone continues to shrink and the ice caps are melting. The world will not change on its own. As founder of the Children's Defense Fund, Marian Wright Edelman stated, "It's time for greatness—not for greed. It's a time for idealism—not ideology. It is a time not just for compassionate words, but compassionate action." If humankind self-destructs, it is not because it lacks opportunities, but because of its inability to come together, spot the problems and to grab them. "Perhaps the common enemy within us will be too strong for us to recognize and overcome," said the astronomer Carl Sagan. He continued, "But, I have hope ... is it possible that we humans are at last coming to our senses and beginning to work together on behalf of the species and the planet?"[34]

I too hope this book may bring people to their senses and come together as one. This book was not meant to predict the future of Earth, but to alter it. That's what I have decided to do with my life and writing this book is just the first step. I could decide to do something else with my time here on Earth, any time I want to, but I won't. I want to create a better world that both my children and my children's children can enjoy. That dedication, more than anything, is what locks in my future and makes it predictable. That aspect of my future is known. For many of you who are reading this right now, especially the younger audience, your future is still undecided. "In the moment that it becomes predictable, I hope you predict it well."[35]

You must remember that the future is cloudy, but we have the power to shape it. Never forget the words of President John F. Kennedy. "However close we sometimes seem to that dark and final abyss, let no man of peace and freedom despair. For he does not stand alone ... Together we shall save our planet or together we shall perish in its flames. Save it we can, and save it we must, and then shall we earn the eternal thanks of mankind."[36]

South African President Thabo Mbeki at the Johannesburg World Summit on Sustainable Development, which took place August 26 to September 4, 2002.

# Notes

**Chapter 1**
**1** Spaulding, Nancy & Samuel N. Namowitz (2005) Earth Science. McDougal Littell: A Houghton Mifflin Company. Evanston Illnois.
**2** Ibid
**3** Ibid
**4** Kegley, Charles & Eugene Wittkopf (2004) World Politics: Trend and Transformation. 9th edition. Thompson Wadsworth. Belmont, CA.
**5** Ibid
**6** Beck, Roger B. *et al.* (2005) World History: Patterns of Interaction. McDougal Littell: A Houghton Mifflin Company. Evanston Illnois.
**7** Ibid
**8** Ibid
**9** Kegley, see prior
**10** Ibid
**11** Ibid
**12** Ponce, Matias (2005, November 4) UNESCO Report Highlights Digtial Divide: New Study suggests 6,000 languages at risk of disappearing. Ohmynews. Retrieved September 15, 2006, from http://english.ohmynews.com/articleview/article_print.asp?menu=c10400&no=256818 ...
**13** Kegley, see prior
**14** Ponce, see prior
**15** Ponce, see prior
**16** Trafton, Anne (2005, November 16) Annan presents prototype $100 laptop at World Summit on Information Society. MIT News office. Retrieved September 15, 2006 from http://web.mit.edu/newsoffice/2005/laptops-1005.html
**17** Ibid
**18** Ibid
**19** Ibid
**20** Bardhan, Pranab (2006, March 26) Does Globalization Help or Hurt the World's Poor? Scientific American. Retrieved September 15, 2006 from http://www.sciam.com/print_version.cfm?articleID=0004B7FD-C4E6-1421-B4E683414B7F0101
**21** Ibid
**22** Ibid
**23** Ibid
**24** Ibid
**25** Beck, Roger B. *et al.* (2005) World History: Patterns of Interaction. McDougal Littell: A Houghton Mifflin Company. Evanston Illnois.
**26** Ibid
**27** Ibid
**28** Kegley, see prior
**29** Jordan, A., Taylor Jr., W., Mazarr, M. (1999) American National Security.

(5th edition).Baltimore and London: The John Hopkins University Press.
**30** Kegley, see prior
**31** Jordan, see prior
**32** Jordan, see prior
**33** Crocker, C., Hampson, F., Aall, P. (2001) Turbulent Peace: The Challenges of Managing International Conflict. Washington D.C.: United States Institute of Peace Press.
**34** Hook, S. & Spanier J. (2000) American Foreign Policy Since World War II. 15th edition. Washington D.C.: Congressional Quarterly Inc.
**35** Ibid
**36** Bush pledges to spread democracy. (2005, January 20) CNN International. Retrieved October 27, 2005 from http://www.cnn.com/2005/ALLPOLITICS/01/20/bush.speech/
**37** Ibid
**38** Dobriansky, Paula. (2004, February 10) Address to Baltimore Foreign Affairs Group. Baltimore, Maryland. [Transcript]. Retrieved October 27, 2005 from http://www.iwar.org.U.K./news-archive/2004/02-10-3.htm
**39** Ibid
**40** United States Agency for International Development. (2005). 2006 Congressional Budget Justification for South Africa. Retrieved on October 27, 2005, from http://www.usaid.gov/policy/budget/cbj2006/afr/za.html
**41** Poll: U.S. Allies Doubt Bush Plan to Spread Democracy. (2005, February 22). VOA News. Retrieved on October 27, 2005, from http://www.voanews.com/english/archive/2005-02/2005-02-22-voa34.cfm?CFID=189111&CFTOKEN=19691424
**42** Hobsbawn, Eric. (2004) Spreading Democracy: The World's Most Dangerous Ideas. Foreign Policy. Retrieved on October 27, 2005 from http://www.foreignpolicy.com/story/cms.php?story_id=2666
**43** Ibid
**44** Mansfield, E. & Synder, J.(1995). Democratization and the Dangers of War. International Security, Vol. 20. No. 1. Reprinted in Brown, M. *et al.* (1999) Theories of War and Peace. Cambridge, Massachusetts: The MIT Press.
**45** Art, R. & Waltz, K. (1999). The Use of Force: Military Power and International Politics. (5th edition). Lanham: Rowman and Littlefield Publishers, Inc.
**46** Gurr, Ted Robert (2001) Managing Conflict in Ethnically Societies: A New Regime Emerges in the 1990's, in Charles W. Kegley Jr. and Eugene Wittkopf (eds.), The Global Agenda, 6th ed. Boston: McGraw Hill.
**47** Ibid
**48** Mansfield, see prior
**49** Kegley, see prior
**50** Bush, see prior
**51** Kegley, see prior
**52** Beck, Roger B. *et al.* (2005) World History: Patterns of Interaction.

McDougal Littell: A Houghton Mifflin Company. Evanston Illnois.
**53** Ibid
**54** Hutton, Patrick (2001) Lecture: Benjamin Barber, Jihad vs. McWorld, an interpretation
Preface: from Western to Global history. University of Vermont. Retrieved September 15, 2006 from http://www.uvm.edu/~hst19/Online_Reading/Lecture_19.htm
**55** Rubin, Barry (2003, September 16) Globalization and the Middle East: Part One. YaleGlobal. Retrieved September 15, 2006 from http://yaleglobal.yale.edu/display.article?id=744
**56** Ibid
**57** Ibid
**58** Ibid
**59** Ibid

**Chapter 2**
**1** Bowman, Michael (2006, March 7) World Population Growth to be Concentrated in Developing Nations, as Total Expected to Reach Nine Billion by 2050. NewsVOA.com. Retrieved October 1, 2006 from
http://www.voanews.com/english/archive/2006-03/2006-03-07-voa1.cfm?renderfor print...
**2** Ibid
**3** Ibid
**4** Ibid
**5** Kegley, Charles & Eugene Wittkopf (2004) World Politics: Trend and Transformation. Ninth edition. Thompson Wadsworth. Belmont, CA.
**6** Ibid
**7** Grain Production Dropping; Fuel Thefts Rising. (2006, July 5) Overpopulation.org. Retrieved October 1, 2006 from http://www.overpopulation.org
**8** Ibid
**9** Ibid
**10** Ibid
**11** Ibid
**12** Population hearings open in U.K. parliament (2006, May 15) People and Planet. Retrieved October 1, 2006 from http://www.peopleandplanet.net/doc.php?id=275
**13** Ibid
**14** Ibid
**15** Ibid
**16** George, Henry (2006) Overpopulation: Is There Such a Thing? Retrieved October 1, 2006 from http://www.henrygeorge.org/popsup.htm
**17** Montaigne, Fen (2002, September) Water Pressure. National Geographic Magazine.
**18** Dugger, Celia W. (2006, August 22) Need for Water Could Double in

50 Years, UN Study Finds. New York Times. Retrieved October 1, 2006 from http://www.globalpolicy.org/socecon/hunger/environment/2006/0822 unwater.htm
**19** Ibid
**20** Ibid
**21** Foulkes, Imogen (2006, August 16) Water shortage 'a global problem'. BBC News. Retrieved October 1, 2006 from http://news.bbc.co.U.K./2/hi/science/nature/4796909.stm
**22** Montaigne, see prior
**23** Montaigne, see prior
**24** People and the Planet (2004, April 30) Population and Human Development—the key connections. Retrieved October 1, 2006 from http://www.peopleandplanet.net/doc.php?id=199&section=2
**25** Populations hearings, see prior
**26** Ibid
**27** Ibid
**28** Bavier, Joe (2006, September 14) 'We've been sick since 1965.' News24.com. Retrieved October 1, 2006 from http://www.news24.com/News24v2/Components/Generic/News24v2_Print_PopUp_...
**29** Ibid
**30** Rincon, Paul (2006, February 20) 'Faster Emergence' for diseases. BBC News. Retrieved October 1, 2006 from http://news.bbc.co.U.K./2/hi/science/nature/4732924.stm
**31** Ibid
**32** Air pollution, cramped living in Athens breeding 'super mosquitoes' (2006, July 4) Yahoo News. Retrieved October 1, 2006 from http://news.yahoo.com/s/afp/20060704/hl_afp/greecehealthdisease_060704163313
**33** Ibid
**34** Ibid
**35** Global Health Reporting (2006) Malaria: State of the Pandemic Retrieved October 1, 2006 from http://www.globalhealthreporting.org/malaria.asp?id=62
**36** Ibid
**37** Ibid
**38** Ibid
**39** Yale Study: Not Enough Metals In Earth To Meet Global Demand (2006, January 27) Yale University. Retrieved October 1, 2006 from http://www.sciencedaily.com/releases/2006/01/060126195628.htm
**40** Ibid
**41** Ibid
**42** Kegley, see prior
**43** Ibid
**44** Ibid
**45** Stalpes, Sean (2006, July 7) Energy Economics For a Sustainable Future:

Curbing our oil dependency, looking at global warming. New West.net: The Voice of the Rocky Mountains. Retrieved October 1, 2006 from http://www.newwest.net/index.php/main/print/10060/
**46** Ibid
**47** Kegley, see prior
**48** Solar Power Generation (2006) Using Solar Power as a source of clean energy. Spaceinfo. Retrieved October 1, 2006 from http://spaceinfo.jaxa.jp/note/kouso/g/kou106_solar_e.jpg
**49** Ibid
**50** Macdonald-Smith, Angela (2006, September 19) China and India trying to harness the power of wind. Bloomberg News. Retrieved October 1, 2006 from http://www.iht.com/articles/2006/09/19/bloomberg/sxwind.php
**51** Stalpes, see prior
**52** Ibid
**53** Siegle, Lucy (January 22, 2006) Is it ethical to have children? The Observer.
Retrieved October 1, 2006 from http://observer.guardian.co.U.K./magazine/story/0,,1691359,00.html
**54** Kegley, see prior
**55** Siegle, see prior
**56** People, see prior
**57** Population hearings, see prior

**Chapter 3**
**1** Pearce, Fred (2006, September 1) Instant Expert: Climate Change. NewScientist.com News Service. Retrieved October 15, 2006 from http://www.newscientist.com/channel/earth/climate-change/dn9903
**2** Kegley, Charles & Eugene Wittkopf (2004) World Politics: Trend and Transformation. Ninth edition. Thompson Wadsworth. Belmont, CA.
**3** European Commission (2005, November 2) Climate Change Briefing. Europaworld. Retrieved October 15, 2006 from http://www.europaworld.org/week211/climate11205.htm
**4** Revkin, Andrew (2007, January) NASA Scientists See New Signs of Global Warming. Retrieved February 2, 2007 from http://environment.about.com/b/a/256856.htm
**5** European Commission, see prior
**6** Matthews, Christopher (2005) Deforestation causes global warming. FAO. Retrieved October 15, 2006 from http://www.fao.org/newsroom/en/news/2006/1000385/index.html
**7** Ibid
**8** French, Hilary (2000) Vanishing Borders: Protecting the Planet in the Age of Globalization. Worldwatch Institute. W.W. Norton & Company: London
**9** Matthews, see prior
**10** Population hearings open in U.K. parliament (2006, May 15) People and Planet. Retrieved October 1, 2006 from http://www.peopleandplanet.net/

doc.php?id=2754
**11** Matthews, see prior
**12** Ibid
**13** Kegley, see prior
**14** United Nations Convention to Combat Desertification (2006) Frequently Asked Questions. Retrieved October 15, 2006 from http://www.unccd.int/publications
**15** Ibid
**16** Ibid
**17** Ibid
**18** Ibid
**19** Ibid
**20** Ibid
**21** Ibid
**22** World Conservation Union (2006, May 6) Release of the 2006 IUCN Red List of Threatened Species reveals ongoing decline of the status of plants and animals. Retrieved October 15, 2006 from http://www.iucn.org/en/news/archive/2006/05/02_pr_red_list_en.htm
**23** Ibid
**24** Ibid
**25** Ibid
**26** Ibid
**27** Amos, Jonathan (2005, May 21) Earth's species feel the squeeze. BBC News. http://news.bbc.co.U.K./2/hi/science/nature/4563499.stm
**28** Ibid
**29** Ibid
**30** Lovell, Jeremy (2006, September 8) Global Warming Taking Earth Back to Dinosaur Era. Reuters. Retrieved October 15, 2006 from http://www.enn.com/today.html?id=11213
**31** Ibid
**32** United Nations Convention to Combat Desertification (2006) Feeling the Heat. Retrieved October 15, 2006 from http://unfccc.int/essential_background/feeling_the_heat/items/2904.php
**33** Ibid
**34** Lovell, see prior
**35** Greenpeace (2006) Our Oceans. Retrieved October 15, 2006 from http://oceans.greenpeace.org/en/our-oceans#
**36** Global Warming International Center (2006, April 20-21) Stormy Weather: Regional Extreme Climatic Swings Occur Globally. Retrieved October 15 from http://globalwarming.net/index.php?option=com_content&task=view&id=130&Itemid=46
**37** United Nations Framework Convention on Climate Change (2006) Feeling the Heat. Retrieved October 15, 2006 from http://unfccc.int/essential_background/feeling_the_heat/items/2904.php
**38** Global Warming International Center, see prior

**39** Ibid
**40** Ibid
**41** United Nations Framework Convention on Climate Change (2006) Feeling the Heat. Retrieved October 15, 2006 from http://unfccc.int/essential_background/feeling_the_heat/items/2905.php
**42** Ibid
**43** Greenpeace, see prior
**44** Ibid
**45** Ibid
**46** Ibid
**47** EU (2005) Winning the battle against climate change. Retrieved October 15, 2006 from http://jpn.cec.eu.int/home/showpage_en_newsobj1008.1.php
**48** United Nations Framework Convention on Climate Change (2006) Feeling the Heat. Retrieved October 15, 2006 from http://unfccc.int/essential_background/feeling_the_heat/items/2905.php
**49** United Nations Framework Convention on Climate Change (2006) Kyoto Protocol. Retrieved October 15, 2006 from http://unfccc.int/essential_background/kyoto_protocol/items/2830.php
**50** Pearce, see prior
**51** Hanley, Charles J. (2006, November 6) Climate Conference Convenes in Kenya. The Associated Press.
**52** Ibid
**53** Ibid
**54** Ibid
**55** Pearce, see prior
**56** United Nations Framework Convention on Climate Change (2006) Feeling the Heat. Retrieved October 15, 2006 from http://unfccc.int/essential_background/feeling_the_heat/items/2905.php
**57** Pearce, see prior

**Chapter 4**
**1** Beck, Roger B. *et al.* (2005) World History: Patterns of Interaction. McDougal Littell: A Houghton Mifflin Company. Evanston Illnois.
**2** United Nations Office on Drugs and Crime (2006) Fact Sheet on Human Trafficking. Retrieved November 1, 2006 from http://www.unodc.org/unodc/en/facts
**3** Ibid
**4** Ibid
**5** U.S. State Department (2006, June 5) Trafficking in Persons Report. Retrieved November 1, 2006 from http://www.state.gov/g/tip/rls/tiprpt/2006/65983.htm
**6** Ibid
**7** Ibid
**8** Beck, Roger B. *et al.* (2005) World History: Patterns of Interaction. McDougal Littell: A Houghton Mifflin Company. Evanston Illnois.

**9** Ibid
**10** U.S. State Department, see prior
**11** Ibid
**12** UNODC, see prior
**13** U.S. State Department, see prior
**14** Ibid
**15** Bush, Laura (2006, September 18) Remarks by the First Lady at the White House Conference on Global Literacy at the New York Public Library. Retrieved November 1, 2006 from http://www.state.gov/r/pa/ei/wh/72382.htm
**16** UNESCO Institute for Statistics (2005) Global Education Digest 2005: Comparing Education Statistics Across the World, Montreal. Retrieved November 1, 2006 from http://www.uis.unesco.org/template/pdf/ged/2005/ged2005_en.pdf
**17** Ibid
**18** Bush, see prior
**19** Ibid
**20** Ibid
**21** UNESCO, see prior
**22** Ibid
**23** Mock, Gregory (2003, July) Undue Influence: Corruption and Natural Resources. World Resources 2002-2004 (Box 2.4, pp. 36-37). Retrieved November 1, 2006 from http://earthtrends.wri.org/text/environmental-governance/feature-44.html
**24** Ibid
**25** Ibid
**26** Ibid
**27** Transparency International (2006) Costs of Corruption in Public Contracting. Retrieved November 1, 2006 from http://www.transparency.org/global_priorities/public_contracting
**28** Ibid
**29** Mock, see prior
**30** Ibid
**31** Transparency International (2006, October 4) Bride Payers Index, BPI 2006. Retrieved November 1, 2006 from http://www.transparency.org/news_room/in_focus/bpi_2006#pr
**32** Ibid
**33** Transparency International (2006, November 6) Corruption Perceptions Index, CPI 2006. Retrieved November 6, 2006 from http://www.transparency.org/news_room/in_focus/cpi_2006/
**34** Ibid
**35** Beck, see prior
**36** Ibid
**37** Kegley, Charles & Eugene Wittkopf (2004) World Politics: Trend and Transformation. Ninth edition. Thompson Wadsworth. Belmont, CA.

**38** Ibid
**39** UNHCR (2006) The State of the World's Refugees: Human Displacement in the New Millennium. Retrieved November 1, 2006 from http://www.unhcr.org/cgi-bin/texis/vtx/template?page=publ&src=static/sowr2006/toceng.htm
**40** Ibid
**41** Ibid
**42** Guellec, Dominique (2006) Brain Drain. Retrieved November 1, 2006 from http://www.scidev.net/dossiers/index.cfm?fuseaction=dossierfulltext&Dossier=10
**43** Ibid
**44** Ibid
**45** Akl, Aida (2006, May 18) Battling Brain Drain. VOA News. Retrieved November 1, 2006 from http://www.voanews.com/english/NewsAnalysis/BrainDrain2006-05-18-voa55.cfm
**46** Guellec, see prior
**47** Crush, Jonathan & Vincent Williams(2005, June 30). International Migration and Development: Dynamics and Challenges in South and Southern Africa. United Nations Expert Group Meeting: Population Division. Retrieved July 29, 2005 fromhttp://www.un.org/esa/population/publications/ittmigdev2005/P05_Crush&Williams.pdf
**48** Ibid
**49** Ibid
**50** DoctorZin (2005, June 12) Hundreds of Women ... Retrieved November 1, 2006 from http://regimechangeiran.blogspot.com/2005_06_12_regimechangeiran_archive.html
**51** Ibid
**52** Ibid
**53** Agence France Presse (2006, October 10) Violence Against World's Women 'pervasive': Report. Yahoo News. Retrieved November 1, 2006 from http://news.yahoo.com/s/afp/20061010/wl_afp/uncrimewomen_061010213221/

**Chapter 5**
**1** Kegley, Charles & Eugene Wittkopf (2004) World Politics: Trend and Transformation. Ninth edition. Thompson Wadsworth. Belmont, CA.
**2** Beck, Roger B. *et al.* (2005) World History: Patterns of Interaction. McDougal Littell: A Houghton Mifflin Company. Evanston Illnois.
**3** Ripley, Amanda (2006, August 21) How much are we willing to take. Time Magazine. Vol. 168, No. 9.
**4** Ibid
**5** Ibid
**6** Ibid
**7** U.S. State Department (2006) Trends in Terrorism Report 2006.

Congressional Research Services. Retrieved December 1, 2006 from http://fpc.state.gov/documents/organization/69479.pdf
**8** Ibid
**9** Ibid
**10** Ibid
**11** Beck, Roger B. *et al.* (2005) World History: Patterns of Interaction. McDougal Littell: A Houghton Mifflin Company. Evanston Illnois.
**12** U.S. State Department, see prior
**13** Associated Press (2006, October 16) Nuclear Chief: More nations 'hedging their bets.' CNN.com. Retrieved December 1, 2006 from http://edition.cnn.com/2006/WORLD/Europe/10/16/nuclear.proliferation.ap/index.html
**14** Ibid
**15** Ibid
**16** Pincus, Walter (2006, October 20) U.S. Plan for New Nuclear Weapons Advances. Washington Post. Retrieved December 1, 2006 from http://www.washingtonpost.com/wp-dyn/content/article/2006/10/19/AR2006101901863_pf.html
**17** The Issues (2006) ControlArms.org, retrieved December 1, 2006 from http://www.controlarms.org/theissues/
**18** Gamba, Virginia (2006) Focus again on the United Nations. PaxAfrica Vol. 3, No. 1. Retrieved December 1, 2006 from, http://www.saferafrica.org/DocumentsCentre/NEPAD/PaxAfrica/Vol3Nr1/PaxAfrica2no1.pdf
**19** The Issues, see prior
**20** Ibid
**21** Control Arms Campaign (2006, October) Arms Without Borders. Retrieved December 1, 2006 from http://www.controlarms.org/documents/ArmsWithoutBorders_Final-21Sept06.pdf
**22** Ibid
**23** Ibid
**24** Ibid
**25** Ibid
**26** Ibid
**27** Ibid
**28** Shah, Anup (2006, March 27) High Military Expenditure in Some Places. Retrieved December 1, 2006 from http://www.globalissues.org/Geopolitics/ArmsTrade/Spending.asp
**29** Ibid
**30** Ibid
**31** David, Leonard (2006, January 11) Beam weapons almost ready for battle. Space.com. Retrieved December 1, 2006 from http://www.msnbc.msn.com/id/10805240
**32** Ibid
**33** Ibid
**34** BBC News (2005, January 23) U.S. plans 'robot troops' for Iraq. Retrieved December 1, 2006 from http://news.bbc.co.U.K./1/hi/world/americas/4199935.stm

**35** Ibid
**36** Ibid
**37** Ibid
**38** Tiron, Roxana (2004, May) Warfare by Remote Control: Security Robot Could Protect Military Bases. Retrieved December 1, 2006 from http://www.nationaldefensemagazine.org/issues/2004/May/Warfare_by.htm
**39** Ibid
**40** Ibid
**41** Ibid
**42** Doomsday Clock (2007) In Wikipedia: The Online Encyclopedia. Retrieved February 2, 2007 from http://en.wikipedia.org/wiki/Doomsday_Clock
**43** Ibid
**44** Kegley, see prior

**Chapter 6**
**1** Mandela, Nelson (2006) In Wikipedia: The Online Encyclopedia. Retrieved August 1, 2006 from http://en.wikipedia.org/wiki/Nelson_Mandela
**2** Make Poverty History (2006) In Wikipedia: The Online Encyclopedia. Retrieved August 1, 2006 from http://en.wikipedia.org/wiki/Make_poverty_history
**3** Environmental Protection Agency (2000, January 7) Global Warming—Actions—Individuals. Retrieved August 1, 2006 from http://yosemite.epa.gov/oar/globalwarming.nsf/content/actionsIndividual.html
**4** Occupation Memorial (2006) HMD 2006 Theme: One Person Can Make a Difference. Retrieved August 1, 2006 from http://www.thisisjersey.com/hmd/2006theme.pdf
**5** Murtha, John (2001, May 5) Murtha Tells Graduates an Individual Can Make a Difference. Retrieved August 1, 2006 from http://www.house.gov/murtha/news/nw010505.htm
**6** Ibid
**7** BBC NEWS (2000, March 20) The UN's Goodwill Hunters. Retrieved August 1, 2006 from http://news.bbc.co.U.K./2/hi/U.K._news/683690.stm
**8** Ibid
**9** Ibid
**10** United Nations Children's Fund (2006) UNICEF People: International Ambassadors. Retrieved August 1, 2006 from http://www.unicef.org/people/people_ambassadors_international.html
**11** BBC. see prior
**12** Miller, Prairie (2001, June) Angelina Jolie on Filling Lara Croft's Shoes and D-size Cups. Retrieved August 1, 2006 from http://www.nyrock.com/interviews/2001/jolie_int.asp
**13** Jolie, Angelina (2003, October) Notes From My Travels. Pocket Books.
**14** Jolie, Angelina (2006) In Wikipedia: The Online Encyclopedia. Retrieved August 1, 2006 from http://en.wikipedia.org/wiki/Angelina_

Jolie#Humanitarian_Work
**15** Ibid
**16** Ibid
**17** Ibid
**18** Ibid
**19** Global AIDS Alliance (2005, November 18) Angelina Jolie Calls on Congress to Fund Global Children's Programs and Eliminate School Fees. Retrieved August 1, 2006 from http://www.globalaidsalliance.org/pressreleases/press111805.cfm
**20** Curry, Ann (2006, May 5) Jolie: I have a lot to be happy about. NBC News. Retrieved August 1, 2006 from http://msnbc.msn.com/id/12578550/
**21** Ibid
**22** Clayton, Jonathan (2002, Oct 14) Feature: Jolie gives refugee girls a shot at school in Kenya. UNHCR. Retrieved August 1, 2006, from http://www.unhcr.org/cgibin/texis/vtx/news/opendoc.htm?tbl=NEWS&id=3daae1974
**23** United Nations Department of Public Information (2006, May 26) UN Messenger of Peace Michael Douglas Features in Public Service Announcement to Draw Attention to Illicit Trade in Small Arms, Light Weapons. UN Press Release. Retrieved August 1, 2006 from http://www.un.org//News/Press/docs/2006/dc3025.doc.htm
**24** Ibid
**25** Ibid
**26** Ibid
**27** United States Institute of Peace (2003, December) Michael Douglas Hosts Film on Child Soldiers. Retrieved August 1, 2006 from http://www.usip.org/peacewatch/2003/12/soldiers.html
**28** Ibid
**29** Ibid
**30** Oprah's Angel Network (2006) About Oprah's Angel Network. Retrieved August 1, 2006, from http://www2.oprah.com/uyl/angel/uyl_angel_about.jhtml
**31** Ibid
**32** Ibid
**33** Ibid
**34** Oprah.com (2006) Oprah's ChristmasKindness South Africa 2002. Retrieved August 1, 2006 from www.oprah.com/presents/2003/christmaskindness/pres_2003_ck_main.jhtml
**35** Alexander, Mary (2005, June 13) Spirit and Hope in SA: Oprah. Retrieved August 1, 2006 from http://www.southafrica.info/ess_info/sa_glance/demographics/oprah-130605.htm
**36** Ibid
**37** Ibid
**38** Ibid
**39** Ibid
**40** Ibid

**41** Bill and Melinda Gates Foundation (2006) In Wikipedia: The Online Encyclopedia. Retrieved August 1, 2006 from http://en.wikipedia.org/wiki/Bill_%26_Melinda_Gates_Foundation
**42** Microsoft PressPass (2006, June 15) Microsoft Announces Plans for July 2008 Transition for Bill Gates. Retrieved August 1, 2006, from http://www.microsoft.com/presspass/press/2006/jun06/06-15CorpNewsPR.mspx
**43** Loomis, Carol J. (2006, June 25) Warren Buffett gives away his fortune. Fortune Magazine. Retrieved August 1, 2006 from http://money.cnn.com/2006/06/25/magazines/fortune/charity1.fortune/index.htm
**44** Loomis, Carol J. (2006, June 25) How Buffets giveaway will work. Fortune Magazine. Retrieved August 1, 2006 from http://money.cnn.com/2006/06/25/magazines/fortune/charity3.fortune/index.htm
**45** Bill and Melinda Gates Foundation (2006) In Wikipedia: The Online Encyclopedia. Retrieved August 1, 2006 from http://en.wikipedia.org/wiki/Bill_%26_Melinda_Gates_Foundation
**46** Ibid
**47** Paulson, Tom (2003, December 8) Gates Foundation out to break the cycle of disease, Seattle Post-Intelligencer. Retrieved August 1, 2006 from http://seattlepi.nwsource.com/globalhealth/151538_global08.html
**48** Bill, see prior
**49** Paulson, see prior
**50** Ibid
**51** Ibid
**52** Byrne, Kevin (2005) Biography: Bono. Retrieved August 1, 2006 from http://www.atu2.com/band/bono
**53** Ibid
**54** Eckstrom, Kevin (2006, February 3) Bono, After Years of Skepticism, Finds Partner in Religion. Religion News Service. Retrieved August 1, 2006 from http://www.atu2.com/news/article.src?ID=4225&Key=&Year=&Cat
**55** Band Aid (2006) In Wikipedia: The Online Encyclopedia. Retrieved August 1, 2006 from http://en.wikipedia.org/wiki/Band_Aid_%28band%29
**56** Ibid
**57** Denny, Charlotte & Ian Black (2002, March 15) U.S. and Europe boost aid to poorest countries. The Guardian. Retrieved August 1, 2006 from http://www.guardian.co.U.K./international/story/0,3604,667739,00.html
**58** DATA (2005) What is DATA? Retrieved August 1, 2006 from http://www.data.org/whydata/
**59** EDUN (2006) EDUN FAQ'S. Retrieved August 1, 2006 from http://www.edun.ie/faq.html
**60** The Humanitarian Youth Culture Study (2006) U2's Bono #1 Humanitarian Hero + Angelina Jolie #3 Behind Gandhi, According to Youth Market Study. Retrieved August 1, 2006 from http://69.93.14.237/entertainment.bono_humanitarian_2006.cfm
**61** Ibid

**62** Ibid
**63** Clinton, Bill (2006) In Wikipedia: The Online Encyclopedia. Retrieved August 1, 2006 from http://en.wikipedia.org/wiki/Bill_Clinton
**64** Ibid
**65** Ibid
**66** Ibid
**67** Clinton Global Initiative (2006) Clinton Global Initiative: Inspiring Change. Retrieved August 1, 2006, from http://www.clintonglobalinitiative.org/home.nsf/pt_our_mission
**68** Conason, Joe (2005, December 1) The Third Term: The Dawning of a Different Sort of Post-Presidency. Esquire Magazine. Retrieved August 1, 2006 from http://www.clintonglobalinitiative.org/home.nsf/nws/na7C7CC5CE3BFD844E852570CB005CA3B0
**69** Ibid
**70** The Carter Center (2005, November 29) About Us: Jimmy Carter—Biography of the 39th President of the United States. Retrieved August 1, 2006 from http://www.cartercenter.org/aboutus/bio2.htm
**71** Ibid
**72** Ibid
**73** Ibid
**74** Carter, Jimmy (2006) In Encyclopaedia Britannica. Retrieved August 1, 2006 from Encyclopaedia Britannica Premium Service: http://www.britannica.com/eb/article-1155
**75** Conason, see prior
**76** Habitat for Humanity (2006) History of Habitat. Retrieved August 1, 2006 from http://www.habitat.org/how/historytext.aspx
**77** Ibid
**78** Ibid
**79** Ibid
**80** International Campaign to Ban Landmines (2005) Jody Williams Long Biography. Retrieved August 1, 2006 from http://www.icbl.org/campaign/ambassadors/jody_williams/bio
**81** Ibid
**82** Kids for Saving Earth (2006) About KSE. Retrieved August 1, 2006 from http://www.kidsforsavingearth.org/kse.htm
**83** Ibid
**84** Ibid
**85** Lundestad, Geir (2001, March 15) The Nobel Peace Prize 1901-2000. Nobel Foundation. Retrieved August 1, 2006 from http://nobelprize.org/nobel_prizes/peace/articles/lundestad-review/index.html
**86** Drosjack, Melissa (2006, May 24) Oprah Fans Determined to Get Her a Nobel Peace Prize. Fox News.com. Retrieved August 1, 2006 from http://www.foxnews.com/story/0,2933,196695,00.html
**87** International Campaign to Ban Landmines (2005) Jody Williams. Retrieved August 1, 2006, from http://www.icbl.org/campaign/ambassadors/

jody_williams/
**88** Engle, Dawn & Ivan Suvanjieff (1995, May 22) An Interview with the Dalai Lama. Retrieved August 1, 2006 from http://www.peacejam.org/pages/laureates_lama/laureates_lama_interview.html

**Chapter 7**
**1** Kegley, Charles & Eugene Wittkopf (2004) World Politics: Trend and Transformation. Ninth edition. Thompson Wadsworth. Belmont, CA.
**2** Ibid
**3** Ibid
**4** Ibid
**5** United Nations Department of Information (2005, May) UN in brief. Retrieved September 15, 2006 from http://www.un.org/Overview/uninbrief_toprint.html
**6** Ibid
**7** Big News Network (2006, October 14)UN formally appoints Ban Ki-moon as new Secretary-General. Retrieved October 15, 2006 from http://story.philippinetimes.com/index.php/ct/9/cid/2411cd3571b4f088/id/0ffaca07d08a7292/cs/1/
**8** Ibid
**9** United, see prior
**10** Ibid
**11** Ibid
**12** Ibid
**13** Ibid
**14** Ibid
**15** Ibid
**16** Ibid
**17** Ibid
**18** The Millennium Development Goals and the United Nations Role (2002, October) Fact Sheet. Retrieved September 15, 2006 from http://www.un.org/millenniumgoals/MDGS_FACTSHEET1.pdf
**19** Ibid
**20** IANS (2006, September 8) Little awareness about Millennium Development Goals, say experts. Retrieved September 15, 2006 from http://www.newkerala.com/news4.php?action=fulnews&id=18884
**21** Ibid
**22** Ibid
**23** Kegley, see prior
**24** United, see prior
**25** International Criminal Court (2006) In Wikipedia: The Online Encyclopedia. Retrieved September 15, 2006 from http://en.wikipedia.org/wiki/International_Criminal_ Court/
**26** Ibid
**27** Ibid

**28** Ibid
**29** Ibid
**30** Ibid
**31** Ibid
**32** Ibid
**33** Lobe, Jim (2004, July 20) House Republicans Fight International Criminal Court. OneWorld. Retrieved September 15, 2006 from http://www.antiwar.com/lobe/?articleid=3058
**34** Ibid
**35** Interpol (2006, February) Fact Sheet COM/FS/2006-02/GI-02. Retrieved September 15, 2006 from www.interpol.int/Public/ICPO/FactSheets/GI02.pdf
**36** Ibid
**37** Ibid
**38** Ibid
**39** Ibid
**40** Ibid
**41** Ibid
**42** Ibid
**43** Kegley, see prior
**44** Organization of American States (2006) The OAS At a Glance. Retrieved September 15, 2006 from http://www.oas.org/key_issues/eng
**45** Ibid
**46** Ibid
**47** Ibid
**48** Kegley, see prior
**49** Paul, James A. (2000, June) NGOs and Global Policy Making. Global Policy Forum. Retrieved September 15, 2006 from http://www.globalpolicy.org/globaliz/index.htm
**50** Ibid
**51** Ibid
**52** Ibid
**53** International Treaties (2006) In Wikipedia: The Online Encyclopedia. Retrieved September 15, 2006 from http://en.wikipedia.org/wiki/International_treatiesotes
**54** Ibid
**55** Ibid
**56** Ibid
**57** UN Treaty Collection (2006) Treaty Event Focus 2006: Crossing Borders. Retrieved September 15, 2006 from http://untreaty.un.org/English/TreatyEvent2006/Treaties/English.pdf
**58** Kegley, see prior
**59** Bhagwati, Jagdish(2004, June). Do Multinational Corporations Hurt Poor Countries? The American Enterprise Online: Politics, Business and Culture. Retrieved September 12, 2005, from http://www.taemag.com/default.asp

**60** Ibid
**61** Ibid
**62** Crabtree, Vexen (2006, Mar 10) Multinational Corporations versus Democracy: The Fight Between Nation States and Commercialism. Retrieved September 15, 2006 from http://www.vexen.co.U.K./countries/multinationals.html
**63** Ibid
**64** Ibid
**65** Barra, Justin(2001, July). Case Study: Kukdong and Nike. Global Eyes: Who is Watching International Business? Retrieved September 12, 2005, from http://www.georgetown.edu/faculty/banchoff/stuff/globalization/mncs/justin.htm
**66** Ibid
**67** Ibid
**68** Ibid
**69** United Nations Committee on Trade and Development (2002). Multinational Corporations in Least Developed Countries. Retrieved September 12, 2005 from http://www.globalpolicy.org/reform/2002/modelun.pdf
**70** Quinlivan, Gary M. (2000, Nov) Multinational Corporations: Myths and Facts. Acton Institute Religion and Liberty. Retrieved September 12, 2005, from http://www.acton.org/publicat/randl/article.php?id=364
**71** Ibid
**72** Crabtree, see prior
**73** Kegley, see prior
**74** Ibid
**75** Lobe, see prior
**76** Crane, David (2006, August 29) Smoke and Mirrors: The Geneva Conventions in the 21st Century. Retrieved September 15, 2006 from http://jurist.law.pitt.edu/forumy/2006/08/smoke-and-mirrors-geneva-conventions.php
**77** Ibid
**78** Inayatullah, Sohail (1999) World Government, Globalization and UN reform: UN futures and structural possibilities of world governance. Retrieved September 1, 2006, from http://www.proutworld.org/ideology/leadership/futunandwg.htm
**79** Ibid

**Chapter 8**

**1** Kegley, Charles & Eugene Wittkopf (2004) World Politics: Trend and Transformation. 9th edition. Thompson Wadsworth. Belmont, CA.
**2** Ibid
**3** Sawin, Elizabeth (2003, February 3) Petitioning a Government That Does Not Yet Exist. Sustainability Institute. Retrieved August 25, 2006 from http://www.worldbeyondborders.org/index.htm

**4** Ibid
**5** Ibid
**6** New World Order (conspiracy) (2006) In Wikipedia: The Online Encyclopedia. Retrieved August 25, 2006, from http://en.wikipedia.org/wiki/New_World_Order_%28conspiracy%29
**7** Even, Louis (1996) In This Age of Plenty: A new concept of economics: Social Credit. Retrieved August 25, 2006, from http://www.michaeljournal.org/plenty51.htm
**8** Ibid
**9** Ibid
**10** Ibid
**11** United Earth (2006) in Memory Alpha Wikipedia: The Online Star Trek Encyclopedia. Retrieved August 25, 2006 from http://memory-alpha.org/en/wiki/United_Earth
**12** Ibid
**13** Ibid
**14** Beowulf, David (1997) The Greatest Science Fiction Film of All Time. Retrieved August 25, 2006 http://www.angry-viking.com/complex.html
**15** Ibid
**16** Ibid
**17** Ibid
**18** International Relations: World Government Theories (2006) In Wikibooks, the open-content textbooks collection. Retrieved August 25, 2006 http://en.wikibooks.org/wiki/International_Relations:World_Government_Theories
**19** Ibid
**20** Ibid
**21** Hui, Sylvia (2006, June 13) Hawking says humans must go into space. Associated Press. Retrieved August 25, 2006, from http://news.yahoo.com/s/ap/20060613/ap_on_sc/stephen_hawking
**22** World Government (2006) In Wikipedia: The Online Encyclopedia. Retrieved August 25, 2006 from http://en.wikipedia.org/wiki/World_government
**23** Ibid
**24** Kegley, see prior
**25** Ibid
**26** Universal House of Justice (1985, October) The Promise of World Peace. Retrieved August 25, 2006 from http://www.worldbeyondborders.org/index.htm
**27** Ibid
**28** Lamb, Henry (1996) Global Governance: Why? How? When? Part 2. Murchison Chair of Free Enterprise College of Engineering. Retrieved August 25, 2006 from http://www.engr.utexas.edu/cofe/governance/
**29** Shevtsov, Jane (2005) World Law and the Roman Senate's Steps. Retrieved August 25, 2006 from http://www.worldbeyondborders.org/index.htm

**30** Yale Center for Environmental Law and Policy Research Program (1998, June 4-5) Strengthening Global Environmental Governance. New York. Retrieved August 25, 2006, from http://www.yale.edu/envirocenter/research/geg/GEMD1report.html
**31** Ibid
**32** World Federalist Movement (2006) WFM's position on World Enviornment Organization. Retrieved August 25, 2006 from http://www.wfm.org/site/index.php/articles/56
**33** Shevtsov, see prior
**34** Shevtsov, Jane (2005) Space and Human Unity. Retrieved August 25, 2006 from http://www.worldbeyondborders.org/index.htm
**35** Ibid
**36** Carl Sagan, Pale Blue Dot (1994)
**37** Crawford, Ian A. (1993) Space, World Government and the 'End of History'. British Interplanetary Society. Retrieved August 25, 2006 from http://www.worldbeyondborders.org/index.htm
**38** NASA changes vision statement: Senator Fear What's Best (2006) Associated Content. Retrieved August 25, 2006 from http://www.associatedcontent.com/article/50693/nasa_changes_vision-statement-budget,html
**39** Ibid
**40** Crawford, see prior
**41** NASA, see prior
**42** Daley, Tad (2003, Sept/Oct) The Real Why: It's time for the space program to stop going around in circles. It's time to go somewhere again. The Futurist Magazine. Retrieved August 25, 2006 from http://www.worldbeyondborders.org/index.htm
**43** Ibid
**44** Hui, see prior
**45** Ibid
**46** Daley, see prior
**48** Shevtsov, Space and Human Unity, see prior
**49** Ibid
**50** Ibid

**Chapter 9**
**1** Talbott, Strobe (1992, July 20) The Birth the Global Nation. Time Magazine. Retrieved September 1, 2006 from http://www.worldbeyondborders.org/index.htm
**2** Streit, Clarence (1939) The American Example. Retrieved September 1, 2006 from http://www.worldbeyondborders.org/americanexample.htm
**3** Ibid
**4** Ibid
**5** Ibid
**6** Ibid

**7** European Union (2006) In Wikipedia: The Online Encyclopedia. Retrieved September 1, 2006 from http://en.wikipedia.org/wiki/EU
**8** Kegley, Charles & Eugene Wittkopf (2004) World Politics: Trend and Transformation. Ninth edition. Thompson Wadsworth. Belmont, CA.
**9** Ibid
**10** Ibid
**11** Ibid
**12** Ibid
**13** What Future for Europe? (2006) In Europa: The Gateway to the EU. Retrieved September 1, 2006 from http://europa.eu/abc/12lessons/index12_en.htm
**14** Inayatullah, Sohail (1999) World Government, Globalization and UN reform: UN futures and structural possibilities of world governance. Retrieved September 1, 2006 from http://www.proutworld.org/ideology/leadership/futunandwg.htm
**15** Ibid
**16** World Net Daily (2003, May 20) Revealed: U.N.'s plan for global governance: WND probe unearths plot for global taxation, gun control, standing army. Whistleblower Magazine. Retrieved September 1, 2006 from http://www.worldnetdaily.com/news/article.asp?ARTICLE_ID=32662
**17** Ibid
**18** Ibid
**19** Inayatullah, see prior
**20** Ibid
**21** Ibid
**22** Mattern, Douglas (2004, November 29) The United Nations at a Fork in the Road. Association of World Citizens. Retrieved September 1, 2006 from http://www.empowertheun.org/forkintheroad.html
**23** Ibid
**24** Kegley, see prior
**25** Talbott, Strobe (1992, July 20) The Birth of the Global Nation. Time Magazine. Retrieved September 1, 2006 from http://www.worldbeyondborders.org/index.htm
**26** Clinton, Bill (1999, October 8) Confrontation or Cooperation? Speech at the International Conference of Federalism, Mont Tremblant, Canada. Retrieved September 1, 2006 from http://www.worldbeyondborders.org/index.htm
**27** Einstein, Albert (1947, October) Open Letter to the General Assembly of the United Nations. Retrieved September 1, 2006 from http://www.seas.columbia.edu/~ah297/un-esa/ws1997-letter-einstein.html

**Chapter 10**

**1** United States Enviornmental Protection Agency (October 1997) Climate Change and Public Health. Retrieved August 7, 2006 from http://yosemite.epa.gov/oar/globalwarming.nsf/UniqueKeyLookup/SHSU5BNNXJ/$File/

ccandpublichealth.pdf
2 Ibid
3 Ibid
4 Ibid
5 Gordon, S.B. (2005) Leaders, Celebrities, Organizations Call for World Government. Retrieved August 7, 2006 from http://www.worldbeyondborders.org/howwemightsbg.htm
6 Ibid
7 Daley, Tad (2003, Sept/Oct) The Real Why: It's time for the space program to stop going around in circles. It's time to go somewhere again. Futurist Magazine. Retrieved August 25, 2006 from http://www.worldbeyondborders.org/index.htm
8 Label Networks Inc. (2006) About Label Networks. Retrieved August 7, 2006 from http://69.93.14.237/aboutus.cfm
9 Label Networks Inc. (2006) The Humanitarian Youth Culture Study by Label Networks Provides Insight on Youth Perceptions to Environmentalism, Humanitarian Causes, + the Future. Retrieved August 7, 2006 from http://69.93.14.237/humanitarian-study-2006.cfm
10 Label Networks Inc. (2006) U2's Bono #1 Humanitarian Hero + Angelina Jolie #3 Behind Gandhi, According to Youth Market Study. Retrieved August 1, 2006 from http://69.93.14.237/entertainment/bono_humanitarian_2006.cfm
11 Label Networks Inc. (2006) The Humanitarian Youth Culture Study by Label Networks Provides Insight on Youth Perceptions to Environmentalism, Humanitarian Causes, + the Future. Retrieved August 7, 2006 from http://69.93.14.237/humanitarian-study-2006.cfm
12 Ibid
13 Ibid
14 Ebisch, Robert (2006, July) Peaceful, Easy Teaching. Delta-Sky Magazine. Retrieved August 7, 2006 from http://www.delta-sky.com/2006_07/JustCauses/index.html
15 Ibid
16 Ibid
17 Ibid
18 Ibid
19 Ibid
20 World Future Council (2006) Values, Vision, Solution. Retrieved August 7, 2006 from http://www.worldfuturecouncil.org/visions-values-solutions.htm
21 Ibid
22 Ibid
23 Ibid
24 World Future Council (2006, May) Summary Report of the Inaugural Meeting of the Founding Council and the Board of Advisors and Donors. August 7, 2006 from http://www.worldfuturecouncil.org/index.htm

**25** Von Uexküll, Jakob (2006) Overview. World Future Council. Retrieved August 7, 2006 from http://www.worldfuturecouncil.org/overview.htm
**26** Ibid
**27** Ibid
**28** World Future Council (2006) Latest News: World Future Council to be launched in Hamburg. Retrieved August 7, 2006 from http://www.worldfuturecouncil.org/index.htm
**29** Ibid
**30** Masciulli, Jospeh (2004, April) From a Culture of Violence to a Culture of Peace. Found in Tom Keating & Andy Knight's Building Sustainable Peace. 1st edition. United Nations University Press/The University of Alberta Press.
**31** Ibid
**32** Kegley, Charles & Eugene Wittkopf (2004) World Politics: Trend and Transformation. Ninth edition. Thompson Wadsworth. Belmont, CA.
**33** Ibid
**34** Ibid
**35** Yudkowsky Eliezer (2003, February 21) 'PREDICTING THE FUTURE'. Speech at NYU's World Futures: The Science of Humanity in the Epoch of Technology Retrieved August 7, 2006 from http://www.singinst.org/presentations/predictingthefuture.html
**36** Kegley, see prior

# Index


Africa 18, 21, 22, 24, 26, 33, 35-38, 40-42, 48, 50-53, 58, 59, 62, 63, 65, 66, 68, 71-73, 80, 86, 91, 94, 95, 99, 100, 103, 104, 151, 159, 168, 174, 176, 178-180
Alternative energy 44, 45, 101, 175
Angola 22, 65, 80, 81
Annan, Kofi 19, 92, 101, 112, 120, 132, 157
Apollo 149, 150, 176, 177
Arms trade 78, 80

Barber, Benjamin 29
bin Laden, Osama 28, 30, 149
Bono 98-100, 108
Botswana 22, 80
Bush, George W. 23, 25-28, 42, 45, 59, 75, 82, 99, 101, 118, 125, 148
Bush, Laura 63, 64

Canada 27, 35, 68, 70, 71, 73, 77, 120, 159, 160, 161, 168, 182
Carter, Jimmy 100, 102-104, 108
China 18, 33, 35-37, 45, 51, 56, 64, 67, 68, 70, 77, 92, 94, 101, 112, 145, 159
Civil war 22, 39, 79, 90
Climate change 38, 40, 48, 51, 53, 55, 58, 59, 84, 101, 102, 115, 125, 126, 127, 133
Clinton, William J. 25, 26, 82, 100-102, 104, 107, 118, 159
Cold War 25, 69, 77, 78, 80, 116, 156
Computers 18, 19, 57, 83, 88
Corruption 27, 50, 66, 67, 123, 127, 181

Dalai Lama 109, 176, 180
DATA 99, 178
Deforestation 40, 50-52
Desertification 51, 52, 58
Democracy 23-28, 86, 122-124, 150, 154, 159, 160, 162
Digital divide 19, 20

Earth 16-19, 21, 23, 44, 47, 49-51, 58, 83, 103, 107, 109, 115, 135, 136, 138, 139, 141, 143-145, 147-150, 158, 166, 173, 175, 177, 179, 183, 184, 186
Education 19, 26, 41, 60, 62-65, 71, 72, 81, 91, 94, 98, 100, 104, 106, 111, 114-116, 140, 143, 161, 162, 177, 180
Einstein, Albert 141, 142, 168, 171, 185
Europe 33, 42, 49, 51, 52, 55, 59, 71, 73, 75, 81, 111, 119, 121, 132, 141, 154, 155, 167, 170, 174
European Union 49, 73, 111, 122, 131, 152-155, 158, 167

Federalism 135, 152, 158-168
Food 22, 28, 30, 34, 35-37, 39, 40, 47, 50, 51, 59, 61, 64, 69, 95, 104, 112, 114, 124, 128, 129, 142, 174
Foundations 96-98, 101, 102, 108, 132, 137, 177
Fuller, Millard & Linda 105, 106

Gates, Bill 96-99, 177
Geneva Convention 126, 132
Germany 27, 36, 45, 68, 75, 77, 99, 124, 140, 183
Globalization 17, 18, 21, 28-31, 70, 78, 102, 111, 124, 127, 132, 183
Gore, Al 100, 144
Greece 40, 169

Habitat for Humanity 94, 104-106
HIV/AIDS 26, 36, 38, 40-43, 61, 63, 87, 91, 95, 97, 99-102, 108, 113, 115, 127, 133, 175, 177-179
Hill, Clinton 107
Hubble Space Telescope 17
Human trafficking 60, 62, 63
Hurricane Katrina 48, 57, 94, 101
Hussein, Saddam 23

Illuminati 137, 138
Immigration 33, 111, 179
India 22, 33-35, 37, 61, 67, 68, 70, 77, 79, 94, 109, 116
Intergovernmental organizations 110
International Criminal Court (ICC) 118-120, 127, 132, 133, 156, 185
International law 92, 107, 125, 129, 130, 133, 176, 119
International Monetary Fund (IMF) 24, 99, 111, 114, 116
International Space Station 135, 146, 158
International treaties 58, 106, 113, 169
Internet 18-21, 28, 93, 102, 141, 165, 179
Interpol 120, 121, 185
Iraq 23, 66, 67, 79-83, 115, 132, 133, 137, 174
Iran 29, 30, 64, 72, 73, 77, 115
Islam 30, 31, 76, 174

Israel 56, 67, 68, 73, 77, 79, 112, 119, 174
IUCN Red List 52, 53

Japan 27, 59, 67, 68, 76, 77, 81, 106
Jihad 29, 30
Jolie, Angelina 89-92, 100, 102, 108

Kant, Immanuel 140, 141
Kennedy, John F. 147, 158, 161, 186
Kids for Saving Earth (KSE) 107
King Jr., Martin Luther 16
Kyoto Protocol 58, 125

Label Networks 100, 178
Landmines 106, 113, 124
Lasers 44, 45, 82, 83, 127

Malaria 40, 41, 115
Mandela, Nelson 48, 60, 86, 87, 95, 108
Mars 135, 148, 149, 175, 177
Media 19, 24, 28, 29, 74, 75, 116, 140
Middle East 30, 37, 42, 44, 64, 73, 80, 104, 115, 136
Migration 68, 70, 71, 133
Military 69, 74, 80-83, 111, 112, 117, 120, 122, 126, 139, 143, 154, 155, 157, 169, 170, 171, 174
Millennium Development Goals 20, 115
Moon 147, 135, 148, 150, 172, 176, 177
Moon, Ban ki- 112, 113
Multinational corporations (MNCs) 21, 29, 127-131

NASA 39, 49, 81, 135, 147, 148, 177, 179
NATO 111, 121, 141, 154, 122
New World Order 25, 136-138, 175
Nobel, Alfred 107, 108
Nobel Peace Prize 107-109, 180, 181
Nobel Laureates 175
Nongovernmental organizations 145

Oil 23, 34, 38, 43, 44, 50, 56, 66, 111, 124, 137, 139, 174, 177
OECD 24, 43, 71, 116
Organization for American States (OAS) 122, 123
OPEC

PeaceJam 180, 181
Persian Gulf War 23, 69
Population 19, 28, 31-36, 38-40, 43, 44, 46-48, 50, 53, 56, 59, 69, 70, 84, 88, 92, 124, 157, 173

Refugees 25, 40, 69, 70, 89, 90, 111, 113, 126, 136, 185
Robots 82-84
Russia 33, 43, 67, 68, 77, 79, 90, 112, 171

Sagan, Carl 33, 147, 185,
September 11th 31, 44, 63, 75, 137, 149
Small Arms & Light Weapons (SALWs) 79, 92
South Africa 18, 26, 48, 63, 66, 68, 71-73, 86, 94, 95, 123, 176, 180
South Korea 27, 68, 77, 112
Sovereignty 115, 132, 133, 145, 157, 167-169
Space 17, 18, 39, 44, 45, 49, 55, 83, 111, 126, 131, 135, 136, 139, 141, 142, 144, 146-150, 158, 165, 177, 182, 184
*Star Trek* 138, 139
Sudan 28, 47, 65, 69, 79, 80, 90, 103, 119

Tennyson, Lord Alfred 141
Terrorism 26, 27, 31, 74-76, 102, 113, 119, 120, 123, 137, 142, 174, 176-178, 185
Transparency International (TI) 66, 67

United Nations 19, 20, 23, 25, 37, 39, 42, 47, 51, 58, 60, 73, 77, 88, 91, 92, 93, 101, 107, 110-118, 120-125, 127, 129-133, 138, 140, 141, 143, 145, 152, 155-158, 167-171, 175, 176, 185
 Charter 92, 112, 113, 115, 116, 122, 125, 156
 FAO 35-37, 50, 51, 114
 Peacekeeping 112, 113, 115-117, 122, 143, 157
 UNDP 18, 88
 UNESCO 18, 88, 114
 UNFCCC 51, 125-127
 UNHCR 69, 89
 UNIDO 114, 131
 UNICEF 62, 88, 89
 UNIPCC 48
 UNODC 61, 131
 UNWTO 111, 115
 WFP 36
 WHO 41, 111, 114
United States 18, 22, 23, 25-28, 30, 32, 33, 35, 39, 41, 43-45, 47, 49, 51, 52, 56, 57, 59, 61-63, 67, 69-71, 73, 75, 76, 77, 80-84, 87, 88, 90, 91, 92, 95, 99, 100, 101, 105, 106, 113, 118, 119, 124,

128, 130, 132, 136-140, 143, 144, 146, 147, 152, 154, 158-164, 168, 170, 171, 173-176, 179, 182, 184,
Constitution 73, 152, 153, 159, 160, 163, 176
Department of State 25, 61, 62, 75, 76
United Kingdom 27, 53, 54, 68, 71, 77, 87, 91, 112, 129, 136, 160, 170

Weapons of Mass Destruction (WMDs) 76, 84
Williams, Jody 106, 108, 124
Winfrey, Oprah 93-95. 98, 108
Women's Rights 72, 73, 124
World Bank 20, 24, 72, 115, 116
World Future Council (WFC) 142, 181-184
World War II 60, 110, 113, 118, 127, 137, 161
World Wildlife Federation (WWF) 38

# Photo credits

**Chapter 1**

Page 16, Earth/Moon, NASA, http://web.pdx.edu/~tothm/Manuscripts/decline_of_necessity.htm
Page 18, Radio Telescope, http://gtalumni.org/Publications/magazine/sum93/images/limb2.gif
Page 18, CPU Room, photo by Scott Firsing
Page 20 and 21, 3 Graphs, done by Rene Firsing, based on information from http://www.bridgethedigitaldivide.com/dd_latin.htm
Page 21, Tanker, http://www.imageafter.com/image.php? image=blship behind001.jpg
Page 22, Mumbai, photo by Jill and Roy
Page 22, Shanty town in Manila, photo by Mike Gonzalez, http://www.en.wikipedia.org/wiki/Image:Manila_shanty.jpg
Page 23, Bush, http://commons.wikipedia.org/wiki/Image:Bush_UN_.jpg
Page 25, Freedom House Graph, http://pubs.wri.org/pubs_content_text.cfm?ContentID=1739
Page 26, Embassy, photo by Scott Firsing
Page 28, bin Laden, FBI, http://commons.wikimedia.org/wiki/Image:Laden.gif
Page 29, McDonalds, http://commons.wikimedia.org/wiki/Image: McDonalds %27s-shanghai.Jpg
Page 29, Flag Burn/Iran, Used with Permission, http://www.theodora.com/flags/new11/flag_burning_7.html

**Chapter 2**

Page 32, People, http://www.bigfoto.com/africa/ghana/ghana-41.jpg
Page 34, Chickens, http://www.agrabilityproject.org/images/clip_image002_0015.jpg
Page 36, Child Sideways, CDC, http://upload.wikimedia.org/wikipedia/commons/4/47/starved_girl.jpg
Page 36, Single Child, http://www.worldvision.org
Page 37, Thailand Water, Greenpeace, Sataporn Thongma, http://www.greenpeace.org
Page 38, Dubai, photo by Aymx, http://en.wikipedia.org/wiki/Image:Burjjum.jpg
Page 41, Yellow Fever, http://commons.wikimedia.org/wiki/Image:Yellow FeverVirus.jpg
Page 41, Mosquito, Photo by Flikayttgia Kumpak, http://commons.wikimedia.org/wiki/Image: Mosquito_closeup.jpg
Page 41, Malaria Community, Photo by Suprotik Basu, http://images.google.co.za/url?q=http://web.worldbank.org/WBSITE/EXTERNAL/NEWS/contentMDK:20461374~pagePK:64257043~piPK:437376~theSitePK:4607,00.html
Page 42, AIDS Statistics, http://www.globalhealthreporting.org/diseaseinfo.asp?id=23
Page 43, Coal Station, http://www.greenpeace.org.U.K./.../Live/Image/6332.jpg
Page 44, Wind Machines, http://angeles.sierraclub.org/camera/windmach.jpg
Page 45, New Solar Power, http://spaceinfo.jaxa.jp/note/kouso/g/kou106_solar_e.jpg

Page 46, Wind Power Europe 2005, Permission granted by Bruce Douglas, Chief Operating Officer, European Wind Energy Association, http://www.ewea.org/fileadmin/ewea_documents/documents/publications/statistics/2005statistics.pdf
Page 46, Scott Firsing and Grace DeMattia, photo by Kelly DeMattia

**Chapter 3**

Page 48, Katrina Satellite, http://www.spc.noaa.gov/misc/carbin/katrina
Page 49, Traffic, Osvaldo Gago, http://commons.wikimedia.org/wiki/Image:Auto Stopped_highway.jpg
Page 49, Ozone Holes, http://www.nasa.gov/multimedia/imagegallery/
Page 50, Polar Bear on Ice, Joel Garlich-Miller, U.S. Fish and Wildlife Service.http://www.arctic.noaa.gov/images/polarbear/schliebe.jpg
Page 50, Trees, http://www.plu.edu/~hoodbs/img/deforestation-2.jpg
Page 51, Aral Sea, http://commons.wikimedia.org/wiki/7/7b/Aralship2.jpg
Page 52, Oryx, http://commons.wikimedia.org/wiki/Image:Oryx.jpg
Page 53, Polar Bear Family, Gene Augustine, U.S. Air Force, http://www.arctic.noaa.gov/images/polarbear/schliebe.jpg
Page 54, Living Planet Graph, WWF, http://news.bbc.co.U.K./2/hi/science/nature/4563499.stm
Page 56, Oil Spill, http://www.imo.org/Safety/mainframe.asp?topic_id=1180
Page 56, Florida Map, http://nicholnl.wcp.muohio.edu/ChemistrySociety/ChemSocietyIndex06.html
Page 57, Katrina, photo by Augustino, FEMA, http://www.photolibrary.fema.gov/photolibrary/photo_details.do?id=14542
Page 58, Global Emissions Graphs, http://jpn.cec.eu.int/home/showpage_en_newsobj1008.1.php
Page 59, Pollution, August 2005, http://commons.wikimedia.org/wiki/Image:HazeAmpangKualaLumpur.jpg

**Chapter 4**

Page 60, Concentration Camp, http://www.annefrank.org/content.asp?pid=345&lid=2USHMM in Washington, courtesy of Marie Ellifritz.
Page 61, Human Trafficking, photo by Kay Chernush, U.S. State Department, http://gvnet/humantrafficking
Page 63, Children in Classroom in Mexico, Courtsey of Julie M. Kinnaird, Assistant Director for Outreach World View. University of North Carolina at Chapel Hill
Page 65, Children in Classroom in Krakow, Courtsey of Julie M. Kinnaird, Assistant Director for Outreach World View. University of North Carolina at Chapel Hill
Page 66, Starving Mother & Child, USAID, http://www.usaid.gov/locations/sub-saharan-africa/sudan/images
Page 68, BPI Chart, Transparency International, http://www.transparency.org/news_room/in_focus/bpi_2006#pr
Page 69, Refugee Camp, European Institute for Research on Mediterranean and Euro-Arab Cooperation, http://www.medea.be/pics/sotindouf.jpg
Page 70, UNHCR Graphs, http://www.unhcr.org/cgi-bin/texis/vtx/template?page=publ&src=static/sowr2006/toceng.htm
Page 71, Airplane, http://commons.wikimedia.org/wiki/Image:Deltalax.jpg

Page 71, German Doctors Graph, http://service.spiegel.de/cache/international/0,1518,grossbild-575805-399537,00.html
Page 72, Union Rally, Photo by Somar, http://commons.wikimedia.org/wiki/Image:8marchrallydhaka%2855%29.jpg

**Chapter 5**
Page 74, Protest, http://www.tldm.org/News4/Hizbullah.htm
Page 76, Mushroom Cloud, http://www.armageddononline.net/images/nuke1.jpg
Page 77, Titan II, http://en.wikipedia.org/wiki/nuclear_weapons.html
Page 78, 2 handguns, photo courtsey of SaferAfrica
Page 82, F117's, http://www.usafnukes.com/picture_page.htm
Page 83, Laser, AFRL, http://www.msnbc.com/id/10805240
Page 84, Unmanned Aircraft, http://www.dodmedia.osd.mil/Assests/2006/Marines/DM-SD-06-02918.jpeg

**Chapter 6**
Page 86, Nelson Mandela Statue, http://commons.wikipedia.org/wiki/Image:Nelson_Mandela.jpg
Page 87, Nelson Mandela, http://www.org.U.K./data/issue29/history.htm
Page 88, Ronaldo, Antonio Cruz, http://commons.wikimedia.org/wiki/Image:Ronaldo.jpg
Page 89, Jolie, http://en.wikipedia.org/wiki/Image:Jolie.jpg
Page 91, Jolie/Rice, http://en.wikipedia.org/wiki/Image:2005_06_15_rice_jolie_600.jpg
Page 92, Douglas, United States Institute of Peace, http://www.usip.org/peacewatch/2003/12/soldiers.html
Page 94, Harpo Sign, http://commons.wikimedia.org/wii/Image:Harpo_studio_sign_in_chicago_ill_usa.jpg
Page 96, Bill Gates in Poland, http://commons.wikimedia.org/wiki/Image:Bill_Gates_in_Poland.jpg from the website of the president of the Republic of Poland, http://www.prezydent.pl
Page 97, Gates'/Buffet, http://www.gatesfoundation.org/Aboutus/RelatedInfo/Discussion-060626.htm
Page 99, Bono/Kohler, http://commons.wikimedia.org/wiki/Image:Bono_U2_with_Horst_k_%c3%B6hler_at_prague_2000_IMF.jpeg
Page 101, Clinton, http://www.clintonfoundation.org
Page 104, 2 Men in Jungle, photo by M. Katabarwa, Carter, The Carter Center, http://www.cartercenter.org
Page 105, Millard Fuller, HFHI, http://www.habitat.org
Page 106, 200,000th house, HFHI, http://www.habitat.org
Page 106, Jody Williams, photo by Japan Campaign to Ban Landmines, http://www.icbl.org/campaign/ambassadors/jody_williams/bio
Page 107, Clinton Hill, http://www.kidsforsavingearth.org/kse.htm
Page 107, Nobel, http://commons.wikimedia.org/wiki/Image:Alfred_Nobel.jpg

**Chapter 7**
Page 110, UN Headquarters, photo by Scott Firsing

Page 113, Ban Ki-moon, http://en.wikipedia.org/wiki/Image:Yudhoyono_BanKiMoon,.jpg
Page 116, UN Peacekeeper, http://commons.wikimedia.org/wiki/Image:Da-ST-96-01247.jpeg
Page 117, DPKO missions, Department of Peacekeeping Operations, http://www.un.org/Depts/dpko/dpko/bnote.htm
Page 117, Size of Forces, Global Policy, http://www.globalpolicy.org/security/peacekpg/data/pcekprs.htm
Page 118, ICC Headquarters, http://commons.wikimedia.org/wiki/Image: Building_of_the_International_Criminal_Court_in_the_Hague.jpg
Page 119, ICC Map, Wikipedia, Licensed under the Creative Commons Germany, http://en.wikipedia.org/wiki/Image:World_map_ICC_member_states.png
Page 120, Ronald K. Noble, Copyright Interpol, Used by Permission from Interpol, http://www.interpol.int/Public/Region/Europe/conference/erc34/gallery.asp
Page 122, IGO Flags, Photo by Tatiana Maxim, http://www.osce.org/ec/
Page 123, photo courtsey of SaferAfrica
Page 124, Helicopter/Indonesia, World Food Program, http://www.wfp.org/newsroom/in_brief/asia/indonesia/050201_indonesia.html
Page 125, Putin & Bush, White House photo, http://www.pbs.org/newshour/bb/international/proliferation/countries/russia.html
Page 127, Coca-Cola, http://www.healthgap.org/press_releases/02/063002_AU_PR_COKE_pride.html
Page 129, Factory Workers, http://www.thailabour.org/docs/CodesReport/HealthAndSafety.html
Page 131, UNO City, http://www.unvienna.org/unov/index.html

**Chapter 8**

Page 135, Moon Aglow, NASA, http://www.nasa.gov/multimedia/imagegallery / image_feature_290.html
Page 136, Black Helicopters, http://www.rotaryaction.com/pages/xfilesend.html
Page 137, Pyramid, http://www.michaeljournal.org/plenty51.htm
Page 138, Star Trek, http://en.wikipedia.org/wiki/Federation_of_Planets
Page 140, Kant, http://www.hkbu.edu.hk/~ppp/Kant_gallery.html
Page 142, Churchill, http://www.powerlineblog.com/archives/005975.php
Page 142, Hiroshima, U.S. National Archives, Still Pictures Branch, Subject Files, 'Atomic Bomb.'
Page 144, Ice Cap, NASA, GRACE TEAM, DLR, Ben Holt Sr., http://www.nasa.gov/multimedia/imagegallery/image_feature 576.html
Page 146, Launch, NASA, http://www.nasa.gov/multimedia/imagegallery/image_feature_605.html
Page 149, Mars, NASA, Pat Rawlings, SAIC, http://www.nasa.gov/multimedia/imagegallery/image_feature_261.html

**Chapter 9**

Page 151, Egypt, NASA, http://www.globe-images.com/africa.htm
Page 152, Original Seal, http://commons.wikimedia.org/wiki/Image: Original_Design of_the_Great_Seal_of_the_United_States_%281782%29.jpg

Page 154, EU Map, http://commons.wikimedia.org/wiki/Image: EU_Map.jpg
Page 156, UN Peacekeepers, http://commons.wikimedia.org/wiki/Image:Rwandan Peacekeepers_Return_home_from_Darfur_4oct2005.jpg
Page 158, Space Station, NASA, http://www.nasa.gov/multimedia/imagegallery/image_feature_390.html

**Chapter 10**

Page 172, Moon base, NASA, http://spaceflightnow.com/news/n0406/16commission/
Page 174, Ebola Patient, CDC, Dr. Lyle Conrad http://commons.wikimedia.org/wiki/Image:7042_loves-Ebola-Zaire-CDC_Photo.jpg
Page 174, U.S. soldiers, Army Public Affairs, http://www.army.mil
Page 175, B2 Bomber, Sandia National Laboratories, http://www.brook.edu/fpprojects/nucwcost/b2atest.htm
Page 176, Sodwana Bay, photo by Scott Firsing
Page 179, Mya, photo by Scott Firsing
Page 180, PeaceJam, Photo By Robert Ebisch http://www.delta-sky.com/2006_07/JustCauses/index.html
Page 180, PeaceJam, SA, http://www.peacejam.org/photos.htm
Page 221, Smog, NASA, http://earth.jsc.nasa.gov/sseop/efs/lores.pl?PHOTO=STS092-713-32
Page 182, WFC, http://www.worldfuturecouncil.org
Page 184, Florida, , http://www.globe-images.com/
Page 186, Child Globe, http://republican-news.org/archive/2002/August29/29summ.html

*Note*: Every effort has been made to obtain permission to reproduce copyright material, but there may have been cases where the author has been unable to trace a copyright holder. The author will be happy to correct any omission in future printings.